Sharks in the Arts

This book is the most thorough exploration to date of the many ways in which a wild creature has been absorbed, reimagined and represented across the ages in all of the major art forms. The authors consider not only how the identity of sharks in the natural environment became incorporated into a cultural environment but also how sharks came to be considered the most feared creatures in the open oceans as a consequence of this incorporation. Yet sharks are especially important in helping to maintain a balance that is essential to the health of the oceans.

The book begins with a treatment of the three sharks at the top of global shark-attack files from scientific, economic and environmental perspectives. Subsequent chapters engage with cultural representations of sharks in poetry, drama, art, novels, screenplay adaptations and films. Through an exploration of the ways in which sharks have been represented in human culture through the centuries, this book alerts the global community to the importance of sharks as a common cultural heritage. It aims to change perceptions of sharks so that they can become more revered than feared. The authors of this book argue that an increased understanding of sharks should lead to the development of better strategies for shark and human interactions.

This book will be of great interest to researchers and students of the Environmental Humanities, Cultural History and the Arts. It is also excellent supplementary reading for courses in Zoology and Marine Science.

Vivienne Westbrook is an Adjunct Professor at The University of Western Australia and a member of St John's College, Cambridge, UK. She has authored numerous works of cultural history and is currently focusing on literary representation within the Environmental Humanities.

Shaun Collin is a Professor of Neuroecology within the Oceans Graduate School and Deputy Director of the Oceans Institute at The University of Western Australia. He has published extensively on the subject of how aquatic animals, including sharks, detect and process environmental signals.

Dean Crawford is an Emeritus Professor of English at Vassar College, USA and the author of *Shark* (Reaktion) and *The Lay of the Land* (Viking-Penguin).

Mark Nicholls is a Fellow of St John's College, University of Cambridge, UK. He currently serves as Librarian, Tutor and Supervisor in British History, and was President of the College from 2007 to 2011.

Routledge Environmental Humanities

Series editors: Iain McCalman and Libby Robin

The *Routledge Environmental Humanities* series is an original and inspiring venture recognising that today's world agricultural and water crises, ocean pollution and resource depletion, global warming from greenhouse gases, urban sprawl, overpopulation, food insecurity and environmental justice are all *crises of culture*.

The reality of understanding and finding adaptive solutions to our present and future environmental challenges has shifted the epicentre of environmental studies away from an exclusively scientific and technological framework to one that depends on the human-focused disciplines and ideas of the humanities and allied social sciences.

We thus welcome book proposals from all humanities and social sciences disciplines for an inclusive and interdisciplinary series. We favour manuscripts aimed at an international readership and written in a lively and accessible style. The readership comprises scholars and students from the humanities and social sciences and thoughtful readers concerned about the human dimensions of environmental change.

Sharks in the Arts

From Feared to Revered

Vivienne Westbrook, Shaun Collin, Dean Crawford and Mark Nicholls

Routledge
Taylor & Francis Group
LONDON AND NEW YORK

earthscan
from Routledge

First published 2018
by Routledge

2 Park Square, Milton Park, Abingdon, Oxfordshire OX14 4RN
52 Vanderbilt Avenue, New York, NY 10017

Routledge is an imprint of the Taylor & Francis Group, an informa business

First issued in paperback 2019

British Library Cataloguing-in-Publication Data
A catalogue record for this book is available from the British Library

Library of Congress Cataloging-in-Publication Data
Names: Westbrook, Vivienne, author. | Collin, Shaun P., author. | Crawford, Dean (Professor), author. | Nicholls, Mark, 1959– author.
Title: Sharks in the arts : from feared to revered / Vivienne Westbrook, Shaun Collin, Dean Crawford and Mark Nicholls.
Description: New York : Routledge, 2018. | Series: Routledge environmental humanities | Includes bibliographical references and index.
Identifiers: LCCN 2017057459 (print) | LCCN 2017060454 (ebook) | ISBN 9781315681078 (eBook) | ISBN 9781138929661 (hardback)
Subjects: LCSH: Sharks in art. | Animals and civilization.
Classification: LCC NX650.S555 (ebook) | LCC NX650.S555 W47 2018 (print) | DDC 700/.462—dc23
LC record available at https://lccn.loc.gov/2017057459

ISBN: 978-1-138-92966-1 (hbk)
ISBN: 978-0-367-85583-3 (pbk)

Typeset in Times New Roman
by Apex CoVantage, LLC

Contents

Figures

The figures can be found between pages 79 and 80.

Acknowledgements

We would like to thank the National Science Council of Taiwan for supporting the project 'Sharks in Art' which was hosted by the Australian Research Centre of Excellence for the History of Emotions at The University of Western Australia (UWA) in 2013. The generosity of the Departments of Anatomy Physiology and Human Biology, English and Cultural Studies and the Oceans Institute at UWA has greatly facilitated our research at various stages of the project. We are also grateful to SymbioticA, the Institute for Advanced Studies, the Schools of Anthropology, Engineering, Law and Biological Sciences at UWA for their willingness to create opportunities for sharing perspectives on sharks. We would also like to thank the Master and Fellows of St John's College, the University Library, Cambridge, the British Library, Williams College Library and the Art Library at the Sterling and Francine Clark Art Institute for the use of their rich resources.

Many thanks are due to Linda Washington, Robert Bell, Leila Crawford, Nigel Westbrook, Darra Goldstein and Michael Schmidt (OBE) for their keen editorial insights.

We would also like to thank The University of Western Australia (UWA) for logistical support in the preparation of the manuscript and both UWA and the Western Australian Government for financial support for SPC. We would like to acknowledge Caroline Kerr, Lucille Chapuis, Ryan Kempster, Kara Yopak, Channing Egeberg, Nathan Hart, Laura Ryan, Carl Schmidt, Enrico Gennari and Charlie Huveneers for useful discussions around interactions between humans and sharks and shark mitigation.

Without the hard work and generous cooperation of artists this book would not have been possible. We would especially like to thank Isabel Muscat, Matt Hawksworth and Jim Toomey for their generosity in sharing their photographs, cartoons and stories with us freely, and to Wolfgang Leander who has been an inspiration in the quest to change the public face and fate of sharks.

1 Setting the shark scene

The *OED* defines the shark in the following way: 'Shark, n.1.a. A selachian fish of the sub-order Squali of the order Plagiostomi; in popular language chiefly applied to the large voracious fishes of this suborder, as the genera *Carcharodon, carcharias*, etc . . .' (*OED online*). But sharks mean so much more to us than any single definition could convey. Throughout human civilization sharks have been appropriated, and misappropriated, to satisfy a very wide range of physical, intellectual and emotional needs. Cultural representations of sharks have on the whole been negative, often blurring all shark species into one generic monster (Figure 1.1) and encouraging negative attitudes towards them, with disastrous consequences. The three major predatory species consistently placed at the top of the International Shark Attack File (ISAF) are often targets for negative representation in the media. These three are known popularly as the great white shark (*Carcharodon carcharias*, so called because of its ragged teeth) (Figure 1.2), the tiger shark (*Galeocerdo cuvier*, so called because of the vertical abdominal stripes apparent in adults, although infants are spotted) (Figure 1.3), and the bull shark (*Carcharhinus leucas*, so called because of its stocky form and what has been falsely perceived as an aggressive nature) (Figure 1.4). Here, we explain some of the key misinterpretations of shark behaviour and offer alternatives based on the latest research. In short, we seek to change some of the most common misperceptions of sharks from a scientific point of view and lay the foundation for subsequent chapters that explore a wide range of cultural appropriations of sharks in the arts through the ages.

Representations of sharks in emotional contexts

All cultures create and assign value to representations to stimulate particular kinds of emotion, for the sake of religion, political power, social bonding and entertainment. The most abstract of these may be the simple strokes and dots within circles, otherwise known to us as emoticons, that are an extension of early psychological research into emotional responses to facial expressions. They now form an essential component of global text language. We understand the significance of two dots and a right round bracket as an indication of happiness and two dots and a left round bracket as sadness. These are the kinds of representations that a child can understand (Gosselin and Simard, 1999).

In the process of thinking about the role of sharks in culture, we have begun to realize that the iconic triangle of the teeth and fin of the great white shark, traceable to representations of sharks in ancient cultures, has the force of an emoticon. It is apparently simple to understand and is able to elicit an immediate recognition response. Unfortunately, because of cultural representations of sharks across media, the usual response is fear (Figure 1.5). While the frequently castoff teeth are relatively easy to find on the ocean floor, the extracted jaws of sharks require a lethal confrontation, and as a result they have become symbolic trophies of triumph over fear itself. Whether we approach emotions from Darwinian (as evolved), social constructivist (as taught), or even cognitive perspectives (as managed), what we do know is that emotions can powerfully influence our behaviour; and our behaviour impacts on environments.

We have to remember that the fear of sharks is a cultural construct and that representations of sharks in culture have only a tenuous relationship to actual sharks. Sharks are wild creatures that hunt for their survival across the world's oceans, but, remarkably, the chances of having any encounter with a shark, let alone a negative encounter, are extremely small. Rocks, strong currents, surf, or exposure to pollutants are the actual dangers we confront when we enter an aquatic environment, and yet the fear of sharks persists.

From an early stage of development, 'when we perceive our environment, we rapidly integrate large amounts of incoming stimuli into categories that help to guide our understanding of the world' (Brosch, 2010:377). What is more, fear stimuli are 'attended to, perceived, recognised, remembered and associated with adaptive behavioural output faster or more readily than any other emotional stimulus' (Brosch, 2010:381). While this kind of 'healthy' fear has a role to play in ensuring survival, many fears are unhealthy and debilitating. Emotion needs to be managed and turned to our advantage (De Raad and Kokkonen, 2000:487). The ability to manage emotion in one context is transferable to other contexts, and has, as Anett Gyurak notes, 'been related to higher well-being and socioeconomic status' (Gyurak et al., 2011:408). Crucial to the goal of overcoming our fear of sharks is a raised awareness of all of the stimuli, the visual, aural, and verbal rhetorical strategies that are used to deliberately provoke our fear, often for commercial gain (Figure 1.6).

Sharks are everywhere in our shared global culture and permeate many aspects of our daily lives: economic, political, social and environmental. Though many of us have never seen a real shark going about its daily chores in the ocean, our collective fear of sharks as monsters results in attitudes that endorse irresponsible, even cruel, treatment of them. Imaginary monsters may elicit fear responses that feel real, appropriate, and in need of defensive action (Behrendt, 2010:677). We need to develop strategies for checking our fears when we know that our monsters are not real. The promotion of more positive representations of sharks is one of many strategies that can be used to forge more responsible attitudes towards sharks.

We argue here that the great white shark, although deserving of our awe, has been taken out of its natural context and placed in cultural contexts in which its natural characteristics and behaviours have been greatly exaggerated and distorted to encourage fear and hate. The spectacular 'Air Jaws' acrobatics of the great white are ubiquitous on the internet. These images imply that anyone in the ocean is in danger of being rushed at in that way, yet this is one of many strategies that the great white needs to use in order to capture smart, fast and aggressive seals. The camera images of gaping, toothy maws are usually captured by thrusting cameras at them as sharks bite on prey, more often bait! The teeth of most predators are scary when viewed close-up while feeding, including ours. The dorsal fins of the larger sharks strike terror into us, yet they merely help sharks to steer through the oceans. Cultural representation has built these magnificent creatures into monsters by drawing on iconic sharp and jagged shapes that infants are taught at an early stage of development to avoid. Cultural representations of the great white shark in media as diverse as cartoons through to film and sensationalised news reports have consistently ignored the great white shark's many more fascinating characteristics that ought to provoke curiosity, awe and respect, rather than fear.

We begin with a 1569 broadside relating the first recorded accidental capture of a five-metre shark by mackerel fishermen, and its ensuing fate as food in London's taverns. The shark is described in vivid detail, including this description of its mouth: 'it had a round snout, short head and three layers of teeth in the upper and lower jaws, marvaylous sharpe and very short' (Anon., 1569). Interestingly, there is no sign of horror or fear in this account. The focus of public curiosity is the shark's many fins, which are fully illustrated in its accompanying woodblock print (Figure 1.7).

Captain John Hawkins' sailors, who are cited in the broadside as having been able to name the strange fish for the Londoners, would have seen many sharks on their voyages to the Caribbean. In historical accounts of ships' voyages, captains' logs, surgeons' reports and sailors' diaries, there appears to be limited evidence of the fear of sharks. There are observations and there are tales of struggles with sharks, i.e. of fending them off captured whales and of rescuing natives or sailors from shark-filled seas. There are recipes, and there is advice about which species make the best stew; there are even some very sympathetic narratives about sharks, but very few fearsome accounts of monstrous fish eating their way through boats.

In his account of his *Voyages and Discoveries* around the world, Captain William Dampier relates the ever-present threat that tropical storms presented to sailors and ships. As they approached Antigua in a storm, he observes 'Capt. *Gadbury's* Ship, that lay aground before it came, was by it hurled over to the opposite part of the Harbour, and there thrown dry on the Sand. The Day after the Storm, the Shore was strew'd with Fish of divers sorts, as well great as small; such as Porpoises, Sharks, &c. and Abundance of Sea-Fowls also were destroyed by it' (Dampier, 1931:271–272). Here, the detail of the dolphins and sharks among the dead serves to create a sense of the power of the storm to which men and creatures are equally vulnerable, reminding us that all living things are subject to the forces of nature.

William Anderson, who was an apprentice surgeon, eventually becoming Captain Cook's ship's surgeon on his third journey, notes that on Monday 14 April 1777 Captain Cook came upon a small coral island. There were 'no traces of inhabitants ever having been here . . . It may be observ'd at the same time that the vast number of Sharks seen near the shore would perhaps be a sufficient proof of such a Circumstance; for even these voracious animals seem to be less numerous wherever that universal destroyer man frequents, as if they knew their intrepid rapacity amongst their own kind could avail them little against his dexterity and cunning' (Beaglehole, 1967a:851–2). In this report, the abundance of sharks in the area alerts the sailors to the absence of men on the island. What is especially interesting is Anderson's reflection upon the fact that the rapacity of sharks is no match for that of men; he displays a genuine, if uncommon, empathy for sharks.

Early accounts of sailors being attacked tend to be mere matter-of-fact recordings and not at all the kinds of shark attack reports that we are used to in the modern popular press. Captain Cook notes in his journal of Thursday 20 August 1772, in Porto Praya Bay, Santiago – the largest of the Cape Verde islands – the story of a carpenter who was fixing the boat when he fell overboard and immediately disappeared. Bayly, writing for the *Adventurer,* noted that according to one of the sailors he was: 'a rational fellow creature of a gentle and amiable disposition'; 'His good-natured character, and a kind of serious turn of mind caused him to be regretted *even* among his shipmates . . . Humanity stole a tear from each feeling traveller'. A few days later these feeling travellers witnessed a second casualty: '[they] had the misfortune to lose a man overboard who was immediately seized by the Shirks, & carried down' (Beaglehole, 1961:32). This subsequent account of a man falling overboard and being 'seized' by a shark impacts on the first to render the 'goodnatured' carpenter a likely victim of a shark. The 'feeling' observers are contrasted with the unfeeling sea, and the shark, within a rational, objectively framed account of both. The men of 'feeling' are to be pitied as much as the victims themselves, closing the gap between the real sensory feeling of being swallowed and being swallowed up by emotion.

In a *Resolution* journal entry of Wednesday 23 December 1778, Captain Cook notes only gratitude for some winds that saved the ship from a great swell that threatened to beach them, making no mention of the shark attack that day. We know about it only because several other sailors did note it in their diaries. Edgar, the same day,

> saw three very large Sharks about, the Men & Women Swimming too & from the Ship at the same time. About half-past 11 a large shark attackd A man in the water, as Soon as he saw him Opening his Mouth & turning on his Side to lay hold of his Thigh he struck him on the Head with His Hand. Immediately the shark swam off. The Man who was rather Elderly seem'd very much

> alarm'd for a little time altho many Indians saw this and Made a Hallowing at the time, the[y] do not seem to fear them as they Keptd swimming abt as if there had been None seen.'
>
> (Beaglehole, 1967b:484)

Riou also noted, 'we Hooked the same Shark afterwards, the Indian being on board at the same time who as soon as he saw it, shed tears & appeared very much fright'ned. The Shark when it was hooked was fired at three or four times, at which the Indians (tho' they were Close bye) took no kind of notice' (Ibid.).

What this account tells us is that the man instinctively protected himself first and only later responded emotionally when confronted with the dead shark. It is not explained whether this was because he imagined what his fate might have been that day as he stared into the bloody jaws, or whether he pitied the shark. The sailors seem more surprised by, and interested in, the lack of emotional responses of most of the natives to sharks than the sharks themselves. We might infer that the natives seem to be giving the emotional cues to the sailors for appropriate responses to sharks as a part of the natural environment. We particularly notice that even immediately after this attack, the natives continued to swim like the plentiful fish surrounding the sharks.

The cultural shark (triggering fear)

As soon as we look for evidence of a negative emotional response to sharks in early modern literature we find it in abundance. In Charles Cotton's *Scarronides* (1667) sharks are depicted as 'greedy fishes' that Aeneas asks Jove to protect him from as he finds himself in the middle of the sea in a storm. In William Diaper's *Nereides: or Sea-Eclogues* (1712) sharks are friendly compared to women – 'Fierce Sharks by gentle Usage are reclaim'd/But Female Pride is savage, and untam'd' (55–57) – which actually endorses the negative image of sharks.

A more extended treatment of sharks can be found in Robert Charles Dallas's *Amabel* (1823), a poem based on a true story of a ship that was overturned by a shark at sea. The terror felt at the approach of the shark is indicated through the desperate clutching at descriptions to convey how the 'monster', 'Leviathan', 'glutton shark', 'finny wolf with horrid jaw' toppled the ship. The shark's movement is described in the following, wonderfully vivid, way, as 'lightning thro' the deep sea breaking' (157):

> Emerging half his mountain-hide,
> The monster strikes the trembling tide;
> And as he sidelong quivers through,
> His prey he singles from the crew—
> From off the Cutter's deck is seen
> The human form his jaws between;
>
> (173–178)

Thomas Hood's *Sally Simpkin's Lament* (n.d.) is illustrative of more comic accounts, which still maintain the negative image of sharks while using humour to manage the fear generated by what is an horrific attack. The victim is snatched by a shark from his love thus:

> 'Oh! Sally, sharks do things by halves,
> Yet most completely do!
> A bite in one place seems enough,
> But I've been bit in two.'
>
> (13–16)

The lover explains that he won't be visiting her as a ghost, since his legs are now in Columbia and, 'I have got the whole/Atlantic for my waist' (35–36). This comical treatment of a horrific and fatal encounter with a shark alerts us to a broader range of possibilities for emotive treatments of sharks within art.

In George Alexander Stevens' farce *The French Flogged* (1767) two sailors, Ben and Ned, find themselves having to swim to shore after a night at sea. As Ben finds a spring of fresh water he tries to cheer up his companion with the thought that although they have nothing to eat, at least they managed a night at sea without getting eaten: 'as to eating it is very well we did not give the sharks a supper.' They then begin to sing cheerfully. In Richard Cumberland's *Imposters* (1789), Lord Janus and Polycarp's argument ends with Lord Janus upbraiding Polycarp for his ingratitude: 'Wretch, didn't I take you out of the gripe of poverty, from the very jaws of famine? a starving usher in a country school? promoted you to a share in this great enterprize, furnisht and equipt you for it like a gentleman? – And is your hunger now so ravenous as to whet your teeth against your friend? are you so mere a shark as to eat shark? Shame upon you!' (Cumberland, 1789:78). In the elision of the fish and land sharks, Polycarp becomes equipped with the fish teeth to eat his friend. However, the implication is that in eating Lord Janus, he would be eating another shark. We are beginning to get a sense of eighteenth-century literary treatments of sharks being preoccupied with consumption – of eating or being eaten, and for very good reason. This was the era of the slave traders who travelled the seas often with diseased, dying or dead human cargo that they offloaded, leaving sharks to clean up. In the later drama there is a much wider range of representations of sharks, though the idea of the land shark remains. For instance, in Caryl Churchill's *Serious Money* (1987), within a context of buying and selling stocks, Churchill suggests that perhaps oceanic shark repellent might be one means of effectively repelling land-sharks.

Sharks have been a feature of numerous novels and seafaring stories. R.M. Ballantyne's *The Coral Island* (1858) tells the story of a shark pursuing Jack and Peterkin, initially for their fish, then returning to attack them. The boys escape only when Jack seizes the oar and forces it into the shark's mouth. What is interesting about this account, as opposed to so many others, is that it indicates the fact that the shark was more interested in the fish, and only in Jack and Peterkin because of their association with it. This small detail, which has been endorsed by recent marine research, is often left out of shark tales in order to promote the image of sharks as man-eaters. Indeed, most of even late twentieth-century advertising campaigns and films have relied upon this image.

The advertising industry has readily used sharks as a means of generating a range of emotions that might propel the general public into a buying frenzy. During *Shark Week*, five adverts were placed on the internet in celebration of the contribution of sharks to advertising. Gillette's 'ProGlide' razor advert shows two divers in shark cages in a shark-populated area of ocean. The event is announced with the caption 'The World's Scariest Shave'. As the sharks circle the cages the divers begin to shave, one with a disposable blade and the other with the ProGlide. A voice-over announces, 'Research shows that sharks can sense a drop of blood from a quarter of a mile away. Which razor would you use?' The diver with the disposable blade then cuts himself and footage of a shark suddenly turning and heading for prey is spliced in. Sharks are of course motivated to feed when they sense a drop of fish blood, not human blood. The advert thus reinforces misconceptions about shark behaviour, misconceptions that not only generate fear but endorse fear as the normal response to sharks.

In the John West 'fish series' of fishermen and ferocious animals there is a 'Tuna' advert that features the great white shark. It begins with the caption, 'False Bay, South Africa'. The voice-over begins, 'Deep in the cold dark waters that lash the wild west Coast of Africa, an insatiable hunger grows. A hunger that drives the great white giants of False Bay to extraordinary lengths.' The camera zooms in on a fisherman eating his crackers and tuna dip. He suddenly throws one out, and the shark leaps out of the water and snaps it up in the air like a dog after a treat. The fisherman eats the last cracker himself as the voice-over announces that, 'The best tuna is now in a snack to go.' The power relationship of man and shark is humorously reversed as he swallows his last cracker and the camera zooms out to reveal a sea of sharks approaching. The POV (point of view) then switches so that the watching audience follows the sharks heading for tuna and crackers as the narrator announces: 'Tuna to go – the best', cleverly harnessing the audience to a disyllabic brand name rhyme.

Among the many films that have focused our fear on great white sharks, *Jaws* (1975) remains at the top of the list. *Jaws* represents the culmination of centuries of misrepresentation of sharks, in a wide range of art forms, as a menace to humanity. By way of contrast, shark documentaries are usually designed to stimulate sympathy for sharks by showing sharks either in natural habitats, or being caught up in nets and finned – sometimes both. Early- to mid-twentieth-century training films that were used to prepare the army, navy and air force to repel sharks in the event of being adrift at sea, some of which appear to us now as alarmingly guaranteed to provoke the interest of sharks, are reflective of a prevalent collective fear of sharks. Servicemen were reassured that they were equipped to deal with these 'man-eaters' from their life-rafts, though armed with useless deterrents that included tearing up any handy sheets of paper and submerging their heads while screaming! YouTube is a medium through which personal encounters with sharks can easily be shared. Shark-attack footage often attracts bloggers, who provoke more fear than the accidental bite caught on camera. The International Shark Attack Files (ISAF) contain details of all reported shark attacks in the USA,

Australia and South Africa for research purposes. Understanding how victims and witnesses of shark attacks react to sharks, to the bleeding body and to each other, has the potential to make an enormous contribution to a wide range of sociological and psychological studies. Images of shark attacks are powerful forgers of attitude; we must contextualize them appropriately if we are to avoid vapid sensationalism.

Appropriations of sharks in visual art, more generally, from painting and sculpture, including John Singleton Copley's *Watson and the Shark* (1778), and Damien Hirst's *The Physical Impossibility of Death in the Mind of Someone Living* (1991), through to cartoons, are rich resources for understanding cultural representation. Jim Toomey is one example of an artist who takes an active interest in sharks and conservation. His *Sherman's Lagoon* strip features a lazy great white shark whose adventures in the ocean usefully and humorously alert us to a variety of ecological issues that have emerged in recent years (Figure 1.8).

More humorous cartoons involving sharks have tended to appear in ephemera of birthday cards and political or educational posters. Included in the Cartoon stock.com collection is a cartoon which features a great white shark at a grand piano in a restaurant. The toothsome white pianist is humorously reminiscent of Liberace (1919–1987). Observing the empty tables, the manager complains, 'Is this *Jaws* theme tune the only thing you can play?' – thereby signalling the power of John Williams's score to generate fear. There are numerous representations of this kind, which incorporate sharks into incongruous contexts to create humorous outcomes, often while communicating contentious issues. Such ephemera plays its part in helping to undermine the fear of sharks.

What emerges from our survey of shark accounts and shark culture is that there are clearly two traditions of shark representations, the first being the accounts of individuals who live and work with sharks on a daily basis and who are able to put them in perspective, representing them in the context of the ocean. The second tradition is that of creative representation, through which sharks are turned into objects of fear for entertainment. Sharks have been swimming in the world's oceans for over 450 million years. Within civilizations they have been worshipped, boiled or baked and eaten, their skin and teeth used for tools and weapons, their livers prepared for oils and their fins cut off for soup. We have consistently demonstrated our capacity to overpower these creatures, and yet they remain the biggest source of fear in the open ocean. Although we have enjoyed being afraid of our man-eating monsters in poetry, drama, prose and an array of visual arts and media, this fearsome fiction has migrated to fear of real sharks. Attitudes forged by the fear of sharks has become a serious problem for some shark species that are essential to the health of the oceans. With natural heritage becoming an increasing focus of art and science projects around the world, it is hoped that further delineations between fictional and real sharks will emerge, aiding conservation through collaboration.

Human impact on sharks

Ecocriticism can be defined in the simplest terms as 'a methodological approach to literary and cultural criticism that takes "the environment" as its primary focus'

(Shoba, 2012:446). It has done much to facilitate collaborations between the arts and sciences, forging a deeper understanding of our natural heritage and how to maintain it. More recently the focus within the discipline has shifted to the irreversible impact of people on nature. 'In 2000, the atmospheric chemist Paul Crutzen made the first published use of the term "Anthropocene" as the name of a new geological period we are now entering, a successor to the Holocene period. The defining mark of this new period is that the global ecosystem, the ecosphere, is changing in response to the activity of a single species' (Kerridge, 2013:346). For many, this agency has determined not just the shape of the planet but its ultimate devastation through human agency.

Ecocriticism was founded on issues that were almost exclusively terrestrial. 'Green' is now firmly established as the signifier of political, economic, social, literary and artistic engagement with the environment: 'Much recent critical work on the crossing of landscape and literature focuses on the effect of social and cultural changes on the landscape as a product of human occupation and order' (Cenki, 2006:xx). Cenki argues that landscapes are always in a space of competition for some form of representation or ownership, but the oceans are also subject to competition, even though they are significantly less explored or understood. Since 70 per cent of our planet is blue, it may be time to start thinking about the impact that our daily lives are having on it more seriously (Casey, 2006:10). The cultural representation of the ocean has been criticised as an idyllicisation that masks human investment and destruction: 'the ocean has come to signal leisure, relaxation, controlled adventure, and uncontaminated wildlife experience. In its picturesque visual appearances, the consumerist industries and economic interests behind this market are effectively cloaked: the crystal waters become escapist projection screens for first-contact fantasies – with all their political and ecological implications safely hidden' (Dawes, 2014:67).

The world's vast oceanic spaces have been open to representation across the spectrum of human emotion and in every medium for 'capturing' it that human inventiveness has afforded. Yet 'almost entirely lacking from ecocritical scholarship is a critical approach to the global ocean – that immense, life-creating and biosphere-sustaining body of salt water in which all life came to be, to which some (marine mammals) returned after an evolutionary sojourn on dry land, and into which we continue to pour the detritus of our civilization and our lives: garbage, toxic waste, excreta, and sins of all kinds' (Brayton, 2012:18). The ocean has become a tipping site for human waste: our calling card of shameful destruction. As we explore the misrepresentation of sharks, we understand that their experience epitomises the many ways in which oceanscapes have been misrepresented and abused. However, 'modern environmental approaches that treat nature with naive realism or mobilise it as a moral absolute, unaware or unwilling to accept that it is informed by specific cultural and temporal values, are doomed to fail' (McCalman and Frawley, 2014: Preface).

A fundamental shift in the way that we understand human interactions with blue nature is necessary: 'against the traditional belief in a transcendent essence of a people (a nation, a state, a race or indeed a class)' (Saldanha, 2016:431). In order to move forward we need to be aware of the inherent contradictions of advocating conservation as consumers of the planet. There are inherent contradictions in the way we manage the oceans' resources and we need to 'defamiliarise this habit of thinking "globally", while acting anthropocentrically' (Ball, 2008:540). Ecocriticism may usefully alert us to the ways in which our attitudes to the oceans have been absorbed through centuries of cultural misrepresentation. As a discipline, it aims at 'a move toward a more biocentric world-view, an extension of ethics, a broadening of humans' conception of global community to include non-human life forms and the physical environment' (Shoba, 2012:447).

We can begin by being more alert to the considerable differences between sharks that have to live in our polluted oceans and the culturally constructed creatures we know through representation; only then can we begin to understand the extent of the problems that we have created for them and consider strategies to ensure their future.

Sharks inhabit different geographical territories, live at different depths of the oceans, and present a very wide spectrum of experiences with people. Benthic sharks that live primarily at the bottom of the oceans are unlikely to experience humans directly, although they may still be the victims of human fallout in terms of chemical and material waste. Pelagic sharks that frequently feed near the surface are more likely to encounter humans directly, either during human leisure pursuits or indirectly through human industrial processes. It is specifically the larger sharks at this end of the spectrum that have been perceived as a threat to human pursuits and have consequently become the target of international laws, although 'shark' has often been used as a blanket term to cover all species (Techera and Klein, 2014). Governments have a range of strategies for dealing with shark and human interactions that sometimes bring them into conflict with conservation groups. Protected by some countries, but not others, great white sharks, in particular, are at the centre of controversy. When sharks are perceived to be a pest to fish stocks by Fisheries Management or, indeed, when a shark encounter leads to a human fatality, the stakes are raised (Crossley et al., 2014:155).

There is a great deal that we do not know about most shark species, largely due to their relative inaccessibility and the lack of resources needed to observe them in their natural habitat. The fact that they have often not been deemed a worthy subject of serious research by funding bodies has contributed to the problem and testifies to the scale of ignorance about the crucial role that sharks play in the ecosystem. As with any aspect of human experience that is little understood, myth, rather than knowledge, has generally underpinned cultural representations of sharks, obscuring the extent to which we all depend on them.

One solution to the problem that sharks face has been deemed to be a complete ban on finning by Fisheries Management (Dell'Apa, 2014:158). A global shark-fin ban could easily feed a black market that would increase, rather than decrease their value. Serious value conflicts attach to sharks that urgently need to be addressed. The capture and killing of millions of sharks each year is part of a larger problem that requires a complex suite of strategies (Figure 1.9). While shark finning has been a focus of media attention, sharks are being killed for a much wider variety of medical and industrial uses. Human appropriation of sharks is a practice that has deep roots in the earliest civilizations and it continues through to the present. 'Perhaps partly because animals were seen to be exemplars of health in the Plutarchan tradition, possessing a kind of natural medicine in their bodies, they came to represent a medical resource, what twenty-first-century terminology calls pharmazooticals' (Harvey, 2013:115).

Traditions governing the use of animals are very difficult to change. Shark-fin soup has been an important part of Chinese culture since the tenth century. Initially considered a dish fit for Emperors on account of the difficulty involved in obtaining shark fins, it was later to become an important symbol of wealth or status in a wider range of social contexts, from the business meeting through to the wedding banquet. Sharks are valued for their vitality in dishes and medicines that include it as part of a Chinese cultural symbolism that incorporates most edible, and inedible, animals and plants (Dell'Apa et al., 2014). Powerful traditions that incorporate endangered species constitute a serious problem for countries that have conservation high on their political agendas. Ancient beliefs and traditions about sharks have retained their potency, even though scientists have repeatedly warned against consuming toxins in sharks and conservationists have outlined the consequences of depleting shark numbers (Figure 1.10).

Aside from social factors that impact on shark numbers, economic factors have often driven attitudes towards, and legislation governing, sharks. The Convention of International Trade in Endangered Species (CITES) refused to protect eight species of shark under threat in 2010. 'The data available clearly supported the case for these species to be listed, but resistance from countries with significant commercial fishing interests prevented the inclusion of these species in the CITES Appendices' (Techera and Klein, 2014:311). Sharks need advocates like Techera and Klein to make concealed agendas transparent so that fresh strategies can be developed that align 'Blue' needs with political agendas more satisfactorily. A change of attitude towards sharks is a fundamental requirement for progress in this area.

Changing attitudes to sharks

If we want to change attitudes, we need to understand how they are forged. We need to be alert to the choice of language used in representing 'Blue' nature, and the verbal and visual strategies that are used to selectively reveal and conceal: 'the linguistic representations of nature reveal how we conceptualize, comprehend and treat the natural world as well as how we construct ideas about nature and our relationship to it' (Akcesme, 2013:96). One of the most frequent representations of nature in the public domain is as a wilderness abstracted from human experience and filled with dangerous animals; a description that is often reinforced with threatening images (Akcesme, 2013:105). These representations are not conducive to sympathetic policy design. Akcesme argues that ecotourism has also been slow to register the negative impacts of its trade on nature. South Africa, for instance, is often advertised as a thrilling shark-dive destination as a means of boosting the economy, but without addressing the potentially negative implications for the sharks. It has been suggested that incorporating education about sharks may help to change attitudes towards them (Techera and Klein, 2014:315). If ecotourism can develop educational activities that help us to prepare for, and be responsible and respectful in, wild environments it will be a positive step towards helping sharks. Since shark tourism appears to be on the increase due to campaigns to persuade communities 'that sharks, and their fins, are worth more alive than dead' (Dell'Apa et al., 2014:158), there is hope that even more sympathetic policies, resulting from informed attitudes, may emerge.

Shark-diving and shark cage-diving are ecotourism activities that have become controversial in recent years (Figure 1.11). Sharks are deemed to be 'under the control' of tour operators for a paid display, but sharks are wild creatures and even governments that promote these activities have their concerns about it. A recent review of cage-diving operations in California, Mexico, South Africa, South Australia and New Zealand revealed that although shark-diving was a multimillion-dollar business around the world, the industry had developed faster than the legislation to control it. Where operators were left free from constraints they were caught chumming in close proximity to surfers and swimmers (Santa Cruz and Stewart Island). The review noted the need to regulate shark cage-diving, with a focus on 'minimising the potential for impacts on sharks and the local environment, minimising potential risks to and conflict with other water users, as well as ensuring economic viability and industry harmony' (Bruce, 2015:2). Regardless of how much legislation is eventually allowed to govern cage-diving operators, large great white, tiger and bull sharks are creatures that may be tagged and tracked, but not tamed or contained for paying customers. Operators who depend on sharks being tolerant of us pushing cameras in their faces may need to expand their suite of underwater attractions for their businesses to remain viable. While cage-diving has been pitched as a powerful means by which sharks can be saved, it nevertheless rests on the assumption that if sharks cannot be eaten and cannot be harnessed for entertainment they are difficult to protect.

Living in harmony with sharks, depends on widespread inculcation of a responsible and respectful attitude towards the marine environment more generally. Further scientific research will undoubtedly clarify the role that sharks play in maintaining healthy oceans and promote the best interests of sharks in the development of further oceanic business strategies. Through more cultural research we will become increasingly alert to the myths about sharks created by human societies. It is hoped that unwarranted fear of sharks will in time turn to appreciation and more positive representations of sharks in art. But collaborative engagement of both the academic community and the general public is required to create change: 'without this faith in the power of individuals and groups to effect cultural change, through both citizen participation or "citizen science" and education, one is surely lost, as is the environment' (Shoba, 2012:442).

There are several useful psychological and pedagogical models already available for creating changes in attitude. One model suggests a stepped approach from what we would call the infant-in-ignorance stage, through to the adult-in-action stage: 'for example, an education campaign for children aged 2–6 should follow a design that fits the first step of the model, which is "little awareness." . . . Campaigns for adolescents and adults can reach the sixth and final step of the model, which is "action"' (Dell'Apa et al., 2014:155). Negative representations of sharks in our culture are so plentiful, often intrinsic to our enjoyment of our favourite books, games, movies and art, that it is a real challenge to see sharks in a positive light. However, significant progress may yet be achieved through the introduction of more positive representations of sharks to pre-teen popular culture.

Authors of children's books can make a substantial contribution to shaping attitudes towards sharks and marine environments. There are, of course, a wide range of 'nature' books that are aimed at children from three to eleven years old, a crucial stage of emotional and attitudinal formation (Kidd and Kidd, 1996; Gosselin and Simard, 1999; Borgi and Cirulli, 2015). However, many of these texts provide inappropriate emotional cues, provoking fear and inculcating negative attitudes.

Camilla Bedoyere's *Discover Sharks: Great White Sharks* (2013) is aimed at three- and four-year olds. It begins unpromisingly with the text 'A great white is a shark. It looks scary!' inserted in a full-page picture of a great white shark with jaws agape, flashing its sharp teeth. It invites the child to look at its big mouth and teeth equipped for hunting. While this does make the mistake of cuing a child's fear response to sharks, Bedoyere spends the rest of the book working towards a more positive conclusion. Highlighted by a large yellow star on a photo of great whites swimming in the ocean, a caption reads 'Now we know that great white sharks are amazing animals!' There are many such books aimed at a very young readership which do not sufficiently take into account the fact that these emotional cues shape attitudes that will form the foundations of their responses to sharks in adulthood.

Mary Cerullo's *The Truth About Great White Sharks* (2005), aimed at four- to six-year-olds, is informative and devoid of sensationalism, while not dodging the difficult fact of shark attacks on humans: 'Some scientists think that sharks attack humans by mistake, since surfers on their boogie boards and divers in their wetsuits and flippers look a lot like the great white's favourite meals – sea lions and seals. Shark experts are also trying to figure out why so many people survive an attack' (Cerullo, 2005:28) (Figure 1.12). Cerullo presents shark attack not as something children need to be afraid of but as a subject for further investigation.

The Shark Caller (2016) by Dianne Wolfer is an imaginative foray into the world of the shark callers of Papua New Guinea. The book is an adventure story written for eight- to ten-year-olds. The story focuses on a little girl named Isabelle who must perform a ritual which involves diving with sharks. While in all other respects the book is an admirable introduction to marine life, the depiction of sharks, and more specifically responses to them, is not as helpful. As Isabelle swims, she describes how 'the sharks nudge me. Sandpaper skin scrapes my hand. I open my eyes, terrified to see, but more terrified not to. Massive creatures circle me. Sharks of all shapes and sizes, their razor teeth so close to my face.

A hammerhead eye stares into mine. I stifle a scream' (Wolfer, 2016:224). The reference to the nudging is an accurate description of one of the ways in which sharks investigate objects in their environment, but the scraping of the hand would only happen if the shark was swimming backwards. Shark skin is made up of tiny denticles that facilitate their movement through the water; rough when stroked from back to front, they are smooth when stroked from front to back (Atkinson and Collin, 2012). Wolfer's refererence to 'razor' teeth against the little girl's face, her 'terror' and stifled 'screams' are all emotional cues to child readers to fear sharks when they have no other reason to fear them. Elsewhere, sharks are referred to by a collective pronoun, 'a pack of sharks' (Wolfer, 2016:146), which is a deliberate choice over the correct terms 'shoal' or 'school' in order to associate sharks with similarly unfair cultural representations of wolves or wild dogs as threatening. In other respects, this is an enjoyable narrative about a child's engagement with marine life, but it is a missed opportunity for a more useful exploration of what is truly exciting about sharks.

How we feel about sharks, which is usually the result of cultural representation rather than engagement with real sharks, has real ecological consequences. It is therefore crucial that these negative emotions towards sharks are addressed. Since knowledge is no guarantee of change, a shift from negative to positive emotions and attitudes can only happen where there is a collective will to change. Key to making a difference to the experience of sharks and other threatened species on the planet is comprehensive, cross-cultural co-operation in the way we represent them. This kind of change cannot be legislated, in the way that finning can; it must come from a real desire to create attitudes in the interests, not at the expense, of sharks.

Image making and image breaking

Some of the most provocative incitements to fear have been created by representations of sharks to the public following a shark 'attack'. It is not uncommon for a report to include a picture of a great white shark with its jaws agape, even when the perpetrating shark is of a different species. Images have the power to provoke emotion and consolidate attitudes: 'Hand-in-hand with a political – in its broadest possible sense – investment in the persuasive role of images, we live in an age of iconoclastic practices underwritten by iconoclastic beliefs. Iconoclasm and the destruction of images is everywhere the solution, marker, and fait accompli of social revolution in our age' (Guerin, 2015:2). Iconoclasm is derived from two Greek words, 'εἰκονο' and 'κλάστης', which together mean 'image breaker'. While it has historically been associated with religious or political revolt, it can refer more broadly to the breaking of any symbolic image. We need to break images that have been used through a variety of media to provoke the fear of sharks. The iconic shark monster is a symbolic image, one that we love to fear. Toppling this cultural construct will leave more space for consideration of the much more exciting creature of the ocean. Being able to identify with sharks, especially those that are the Shark Attack File 'stars', knowing more about them and engaging with them, are important steps towards feeling that we want to protect them.

While Frances Guerin's exploration of the politics of traumatic images has a useful contribution to make in terms of how we understand a variety of their functions in culture, her belief in the efficacy of such images to remind us to avert catastrophe in the future is somewhat naive (Guerin, 2015:3). Visual records and representations of shark attack, for instance, do not stop people from entering the water, but they do increase fear and spur governments to 'ensure safety' through a range of often foolish and catastrophic strategies. One recent study measured the effectiveness of information about mitigation strategies on two of Australia's beaches by assessing the extent to which it affected people's attitudes to sharks. It was noted that the role of the media contributed significantly to a collective fear of sharks and as an effective emotional cue: 'most media coverage emphasises the risks that sharks pose to people, with shark attacks being the focus of over half the U.S. and Australian newspaper articles related to sharks (Muter et al., 2013). Media exposure following most shark attacks has been shown to increase the perceived level of risk (Slovic et al., 2004)' (Crossley et al., 2014:161). Crossley's study found that knowledge about mitigation strategies had no impact on which beaches the 766 Australians interviewed chose to swim at and that fear of sharks remained prominent (Figure 1.13).

A recent study was conducted of the responses of 531 American *Shark Week* viewers to a range of shark-attack images with public service announcements by both celebrities and action group representatives. All of the viewers were recruited from Amazon's Mechanical Turk platform. Their aim was to assess how these announcements influenced 'audiences' emotional reactions, threat perceptions, willingness to support conservation, and intentions to seek information' (Myrick and Evans, 2014:544). The focus was specifically on the extent to which *Shark Week* affected the fear of sharks, testing the claim that 'Media have long been recognized as having the ability to scare audiences (Janis and Feshbach, 1953; Witte, 1992; Cantor, 2009)' (Myrick and Evans, 2014:546). Myrick and Evans hypothesized that watching *Shark Week* 'may lead audiences to become overly fearful of being victimized by sharks and/or less willing to make sacrifices in order to protect sharks and their environment' (Myrick and Evans, 2014:547). Their study found that whether the shark attack presented was real, realistic, or highly fictionalised, for instance in cartoon form, viewers' perceptions of risk of attack increased fear. They also found that even when scientific statements are introduced into *Shark Week*, reminding people of the small likelihood of being attacked, Discovery Channel still presented sharks 'as frightening and potentially violent animals in ways that resonate with audiences' (Myrick and Evans, 2014:560).

Findings from both legal (Techera and Klein, 2014) and psychological (Myrick and Evans, 2014) shark studies appear to point to direct or indirect refusal of scientific information, in the first case because of vested interest, and in the second because of negative emotion. Myrick and Evans found that fear, propelled by images of sharks as violent, overrides rational arguments for not being afraid of sharks because of the low probability versus the traumatic impact ratio of shark attack. Systematic representation of sharks in non-violent contexts is a crucial iconoclastic strategy. Public pressure does seem to be the biggest factor in successfully transforming cultural practices. Businesses need consumers and governments need to be elected. Learning to revere rather than fear sharks is a challenge that must be embraced collectively, and across disciplines.

Shark attack or shark engagement?

There is convincing evidence that sharks are directly or indirectly regulating the natural balance of ocean ecosystems, at all levels, and so are an integral part of the future health of over 70 per cent of our planet. One study that exemplifies the imbalance created when sharks are removed comes from the University of Miami. This study revealed how the 14 species of organisms that formed a large part of the diet of 11 eliminated shark species became overpopulated. This created a knock-on effect of increasing the abundance of the major predator of the bay scallop (the cow nose ray). The scallop fishery, which had been thriving sustainably for over 100 years, was virtually wiped out, with scallop catch dropping by nearly 90 per cent. The removal of the scallops also had an effect on water quality as there were no longer enough scallops to filter and clean the water. It has long been known that sharks help to minimise such depletions and ensure the health of the oceans by keeping other predators in check (Myers et al., 2007).

Although we may think we know a lot about sharks, which is why it has been so difficult to change our attitudes to them, the truth is that there is a great deal more to be learned. The extent of the threat of extinction among the estimated 500 species of sharks is a point of contention between conservationists, fisheries scientists and governments around the world (Field et al., 2009; Ferretti et al., 2010; Dudgeon et al., 2012). Key to understanding this important issue is being aware of the fact that mature great white, tiger or bull sharks, although top predators, can take 15 years to mature, or in the case of great white female sharks 33 years (Natanson and Skomel, 2015); they are prey to other creatures until then. An added problem is that these larger sharks do not reproduce regularly, and when they do gestation is a slow process that takes between 11 and 16 months. This means that sharks cannot easily be replaced once fished out.

In an extraordinary article published in the journal *Marine Policy*, Dr Boris Worm suggested that 100 million sharks were being taken from our marine ecosystem every year (Worm et al., 2013), the best estimate we currently have of the kind of pressure that sharks are being subjected to in the world's oceans. They simply cannot replenish themselves at the rate at which they are being fished (Bensley et al., 2010; Werry et al., 2014). The depletion of sharks could seriously upset the balance of trophic food chains. The overexploitation of sharks has reached such high levels that some species are currently at significant risk of irreversible decline, while others are already near extinction. The full extent of susceptibility of population collapse remains hard to predict without further research.

Of the estimated 500 species of shark, the great white, the tiger and the bull shark are the only species that have any significant recorded collisions with human beings in the ISAF, but this does not mean that they are actively pursuing humans as prey. As a recent study has demonstrated, there are many environmental and contextual factors that contribute to shark attack (Chapman and McPhee, 2016). It is important to understand that these same three species are also indicated as being at risk on the International Union for the Conservation of Nature (IUCN) listings.

Numerous studies have demonstrated that shark attacks are very rare, averaging fewer than ten fatalities per year worldwide. Most attacks are recorded as being non-fatal bites to arms and legs, where the shark has either mistaken a human for what it considers to be an edible food item, or has merely bitten the item in its environment out of curiosity. Shark's mouths have high concentrations of taste buds located amongst their many rows of teeth that enable them to make judgements based on one bite (Curtis et al., 2012; Rasch et al., 2016; Atkinson et al., 2016); hence, sharks bite and spit.

While there are scientific studies of various aspects of shark behaviour, much of this science is conducted at a distance from sharks, often in unnatural surroundings, in aquaria or laboratories, and some are conducted post-mortem. Scientists are now exploring the central nervous system or brain of different species of sharks, predicting search and feeding behaviour based on the relative size of the parts of the brain dedicated to sight, smell, hearing and the ability to detect weak electric fields of their prey (Yopak et al., 2010; 2014). Given the sheer size of these large predators and the difficulties in observing their behaviours under natural conditions, it will be some time before we can uncover the behavioural drivers of searching, targeting and making attacks on natural prey, not to mention how they find mates, form social groups and migrate long distances. However, uncovering the neural bases of shark behaviour will ultimately allow us to understand and mitigate negative human-shark interactions (Collin et al., 2015).

It is difficult for scientists to uncover the drivers of behaviour to predict how different species of sharks will react under specific conditions. The environments in which they live are varied, complex, and constantly changing. Observing sharks in their natural habitat takes time, resources and training. Locating sharks in the numbers required to provide evidence of predictable behaviour(s) can be also a problem, especially when some species migrate long distances and aggregate only seasonally. The professional ethics by which shark scientists are governed have also determined the level of access to sharks and the kind of knowledge that can be gained, where both human and shark welfare must always be considered.

Complementary sources of information are the accounts of recreational shark divers who travel across the world to experience interactions with (often large) sharks in their own environment. A few simple strategies that are well used by divers can help to minimize the risk of having any negative encounter with a large shark, such as always remembering to keep our eyes on sharks in our immediate environment and standing our 'ground' when approached at close quarters, calmly allowing them to investigate and move on once they have established that we are neither food nor a threat. Wolfgang Leander is a highly renowned and respected shark diver. He has been free diving with sharks since the 1950s. Although he was initially afraid of large predatory sharks because of what he had understood through sensationalised accounts of them, through his own experience of being with them in their own environment in a non-threatening capacity he soon realized that, contrary to their reputation as senseless killers, they were very shy, curious, highly sensitive and remarkably intelligent animals. His life-long love affair with sharks is well documented (Weisz, 2011). However, his contribution to changing our perceptions of sharks through his breathtaking close-up photographs taken with his small Nikonos V camera deserves to be more publicly acknowledged (Figure 1.14). In the photograph by Isabel Muscat we see what looks initially like a shark attack amusingly become apparent as a relaxed and friendly engagement between Wolfgang and a female tiger shark in the Bahamas (Figure 1.15). More work is beginning to emerge as a result of personal interactions with sharks, more recently by Ila France Porcher (2017), that usefully augments our awareness and understanding of shark behaviour.

The more we talk to divers, the more we learn about how, and how not, to behave in the presence of sharks. Matt Hawksworth, who is well accustomed to diving with great white sharks, observed on one particular expedition a 3.3 metre gentle male great white shark, which the divers named Marshmallow since it was highly interactive with them, allowing them to scratch his head and even playfully tug on its pectoral fins. Unfortunately, when the shark travelled to another boat he received much rougher treatment; the operator floated the cage out and then trapped Marshmallow against the side of the boat. Matt describes how 'after much thrashing Marshmallow got away from the boat and was then very aggressive. He had pressure tears in his flesh from the cage on one side and was covered in anti-fouling paint on the other side of his body. He was no longer the comfortable and calm creature we had enjoyed being with for several days before he was abused' (Hawksworth, 2017).

These informal interactions can be used to establish patterns of behaviour of different shark species within a range of different contexts. Divers who work with sharks in their natural environment on a daily basis, i.e. filming, framing them, and, through various technological processes, representing them, should be considered as citizen scientists and shark experts in their own right. They do not control sharks, but they clearly are mediators between sharks in their natural habitats and the public view, and there is much to be learned from them. 'Emma' has devotees who have snapped her, assigned a birthday to her and given her a Facebook page. Emma is a real shark, subject to the conditions of life in the ocean, but she has become 'open art' within human culture. Recent research conducted while diving with tiger, lemon and reef sharks confirms that they do investigate new phenomena within their territory, but reveals no evidence of them mistaking divers for their usual prey, even when the divers are surrounded by chum and situated very close to a bait crate. By considering divers' stories from a variety of perspectives, it should be possible to more accurately determine what stimulates certain behavioural responses in sharks to human beings and to understand more accurately what constitutes a provoked and an unprovoked 'attack', the two categories used by the ISAF. The relationship between sharks and divers is in flux, but these recorded positive shark engagements create and promote new images of sharks in their own environment. The use of underwater drones and 'critter cam' approaches, as long as they remain non-harassing, can also further our investigations of natural behaviours, the conditions under which sharks move closer towards areas populated by humans, and their responses to mitigation strategies.

Understanding attack

Sharks cue in on sensory stimulation, often over very long distances, and are attracted by signals that are perceived to satisfy their survival instincts. This accounts for the fact that most attacks are not on intrepid surfers over deep water in 'shark territory', but rather on swimmers and surfers within tens of metres of the shoreline. That said, Nick Fanning recently had a close encounter with a great white shark, caught on camera (Russell, 2015). Although this was obviously not an attack, the cues elicited by his presence on a surfboard alerted sharks to approach. These cues could range from perhaps visually resembling a seal (based on either its shape or movement), producing low frequency vibrational and bioacoustic signals that we know sharks are sensitive to, and/or chemical cues. Some of these cues may be perceived by a shark as a prey item that is somehow distressed, while others may elicit an innate predatory reaction that targets healthy prey.

Although, in many cases, the exact conditions under which attacks occur are not known, it appears that many of the attacks on surfers occur when they are sitting on their boards waiting for a wave, dangling and kicking their legs in the water. These low frequency disturbances will be detected by sharks hundreds of metres away and indicate a potential opportunity to feed. As sharks approach, other cues (such as visual) will become more important, depending on the prevailing environmental conditions (such as water clarity), the sensory capabilities of the species and its motivation. The drivers of behaviour may even be different for individuals of the same species. They appear to have their own 'personalities', with younger, often juvenile, individuals generally displaying more aggressively inquisitive behaviour towards potential prey items. These variables emphasize the difficulty in predicting behaviour under all conditions. A great white propelling itself at 35km per hour from depth towards a surfer that it perceives as a turtle or a marine mammal may be unable to avoid a collision, often with fatal repercussions. There are many examples of attacks that involve a single bite, after which sharks have immediately left the area. Furthermore, sharks have been observed biting a large range of objects in the water that they have subsequently rejected as food (Hammerschlag, et al., 2012). The ISAF files are a useful resource for understanding shark behaviour leading up to and following an attack, within the context of trauma.

Unlike rays and their close relatives the elephant sharks, which have a well-developed colour vision system, sharks are generally colour blind (Hart et al., 2011). Although not all species have been examined, it appears that this group of cartilaginous fishes do not possess the retinal machinery to process colour and instead view their environment in shades of green. Therefore, despite living in an environment predominated by blue light, these apex predators are unable to perceive short wavelengths, relying on contrast to visualize prey, predator and conspecifics. For great white, tiger and bull sharks, surfers or swimmers at the surface, which may cast a high contrast silhouette under bright light conditions in relatively clear water, could be mistaken for a large turtle or seal (Hart et al., 2011). If the approaching sharks also perceive objects to be similar in shape and size to a familiar prey item that moves in the same way and produces an acoustic disturbance, they will probably approach closer to investigate. It is interesting to note that even larger apex predators, such as many marine mammals (whales, dolphins and seals), are also cone monochromats, or colour blind, and that contrast rather than colour plays a major role in the detection and discrimination of predators, prey and conspecifics (Peichl et al., 2001; Newman and Robinson, 2005). This basic knowledge of how large marine predators perceive their environment may enable more effective strategies to be developed, at least for those sharks for which vision plays a major role, so that attacks on humans may be avoided.

Early testing of what was called the 'Shark Screen' by the US Navy consisted of a large, impermeable bag with three inflatable collars around the opening. It was designed to provide physical protection against a shark by completely concealing the occupant visually, containing any body odours or effusions (e.g., blood, urine) that might attract sharks, isolating any bio-electrical emissions, and dampening any water displacements caused by body movements. Tests with captive and wild sharks revealed a general reluctance by sharks to approach the bags, even when motivated by the presence of food (Tester et al., 1968). The data also indicated that white and silver bags with high reflectivity were more attractive to sharks, while blue and black bags of low reflectivity were significantly less attractive (Gilbert, 1970). Subsequent testing with other sea-survival equipment, including yellow lifejackets and a cocoon-like infant flotation device (IFD) revealed that sharks were readily attracted to, and attacked, such objects (McFadden and Johnson, 1978), giving rise to the term 'yum yum yellow'. Our current knowledge on the importance of contrast in the visual world of sharks is now able to explain some of these previous findings as resulting from high contrast rather than colour *per se*.

The same principles of visual camouflage may apply to swimming apparel such as wetsuits. With this in mind, reflectance spectra that provide minimal visual contrast against the water background for the shark visual system have been calculated and used to design 'cryptic' wetsuit patterns that may reduce the risk of shark bite (SAMS, 2013). Field testing of the wetsuit prototypes is still required to determine whether a camouflaged design is less attractive to sharks than a standard black wetsuit. A black-and-white banded wetsuit that is intended to mimic the banded poisonous sea snakes that many sharks appear to avoid (Nelson, 1983) has also been developed based on presenting a conspicuous, but aversive, pattern that takes into account the visual acuity of the shark visual system (SAMS, 2013). However, it is worth noting that tiger sharks are not deterred by banded sea snakes, underscoring our need to understand that each species of shark perceives its environment differently.

Sharks, like many other wild creatures, must hunt for food without wasting energy or putting themselves unnecessarily at risk, and they have specific strategies to help them to do this. All sharks possess a range of senses that allow them to react to a range of environmental cues. Of those species that have been examined, it is clear that the relative importance of each sensory modality is different for each (Collin, 2012; Collin et al., 2015). This is especially important to understand, since sharks are often discussed in the media as though they were all one species, and are sadly all consequently affected by negative publicity. Even for the three species of shark posing the greatest perceived 'threat' to humans, the parts of the brain dedicated to each sense vary enormously. While the visual centres of the brain are hypertrophied in the great white, the olfactory and the electroreceptive parts of the brain are larger in the tiger and bull, respectively. This means that the way in which each of these species processes environmental information, which ultimately guides their behaviour, is different. Great whites appear to be visual specialists, targeting potential food based on contrast, rather than colour (Hart et al., 2011). Anything that appears within a great white's upper visual field that casts a high-contrast silhouette in the shape of a familiar prey object is likely to focus its attention and could elicit an approach.

The success of any visual hunter is, in part, determined by environmental factors (Collin and Hart, 2015). Those species inhabiting clear water bathed in plenty of sunlight possess eyes adapted for diurnal (or daylight) vision with the ability to discriminate fine detail, while species inhabiting deeper water or turbid environments, and therefore dim light environments, possess eyes adapted for nocturnal (or night) vision with higher sensitivity (Lisney et al., 2012). However, in visually compromised environments, sharks often rely on other senses such as olfaction (smell) and/or electroreception (the ability to passively detect weak electric fields). Because the part of the brain dedicated to olfaction is large in tiger sharks, they can sense minute concentrations of odours over hundreds of metres, or even thousands when aided by currents (Yopak et al., 2014). Bull sharks, which live in turbid environments and can migrate between fresh water and salt water, possess high densities of electroreceptors, which allow them to detect and react to the weak electric fields of living animals nearby in the absence of any other sensory information (Collin and Whitehead, 2004; Kempster et al., 2015). This innate sensitivity to electric fields by sharks was first recognised in 1935, when it was shown that blindfolded small-spotted catsharks would react to the presence of a rusty steel wire when brought in close contact with the head. The galvanic currents generated on the wire's surface were large enough to be detected by the shark's very sensitive electrosensory system, resulting in a behavioural response. Subsequent research demonstrated that sharks can be attracted to electrical fields, a behaviour that allows them to find food, but they can also be repelled when an electric stimulus differs in frequency or strength from the bioelectric fields produced by their prey (Kalmijn, 1971). Over 50 years later, the Shark Shield™ – a personal shark deterrent device that produces a strong electric field that overstimulates the shark electroreceptors – was found to be an effective deterrent. Even great white sharks are repelled to a distance of 1.3 m from the device's antenna (Kempster et al., 2016). However, there is a consensus that no deterrent is likely to be 100 per cent effective and that, due to differences in the sensory thresholds of each species, the chances of reducing the risk of an attack may lie in some sort of multisensory device that can affect more than one sensory modality across a variety of environmental conditions.

Given what we know, divers, swimmers or surfers can be in danger when the water has been disturbed and becomes turbid by storm activity, or visually 'polluted' by terrestrial run-off, reducing visibility in what is usually clear water. Such run-off, along with sewage pollution, can also excite olfactory hunters (bull sharks), which are fairly common at the mouths of tributaries in Central America and the northern parts of South America.

Such factors can increase the chances of accidental attacks by sharks on humans. Sporadic changes in environmental conditions not only make it more difficult for humans to see approaching sharks, which is often crucial in deterring inquisitive sharks, but also make it difficult for sharks to navigate and visualize their preferred prey.

The Taronga Conservation Society of Australia incorporates the Australian Shark Attack File (ASAF), a comprehensive database on shark attacks and encounters. It cites all unprovoked cases of shark attack since 1791. Since the earliest recorded incidents, there have been 186 unprovoked shark fatalities in Australia. In the last 50 years, with many more people in the water than ever before, there have been only 47 unprovoked shark attack fatalities, which averages fewer than one per year (0.94). Popular statistics that are often recycled are of the likelihood of being killed by a peanut or a car accident compared with being attacked by a shark (Carey, 2005). The micromort concept (unit of risk, measuring a one-in-a-million probability of death) introduced by Ronald A. Howard helps us to understand that the likelihood of drowning in a marine environment is significantly greater than dying from a shark bite (Howard, 1984).

However, public fear perpetuated by adverse publicity about sharks is likely to render any logical argument based on statistics redundant. What shark-attack files do not tell us is that we still do not fully understand what constitutes a provocation to a shark. Neff and Hueter argue that one reason for a hyped fear of sharks is because of the inherently flawed means by which shark interactions are reported and categorised. They consider 'the dominant perception of intent-laden shark "attacks" with fatal outcomes is outdated as a generic term and misleading to the public' (Neff and Hueter, 2013:65). Indeed, the ASAF tells us that of the 555 cases of 'shark attack' that have occurred over the last 100 years, 102 'attacks' did not result in any injuries to humans!

It is important to emphasize that our fear of sharks is the real problem, not sharks themselves, and that this fear can have a major detrimental impact on the economies of countries for which tourism is a major industry (Kock et al., 2010). Public perceptions of sharks and ineffectiveness of protection against attack become negatively reinforced, leading to the often also unpopular 'introduction of culling programs by authorities and illegal hunting' practices (Kock et al., 2010:448). Nothing can be done to protect swimmers who choose the ocean over a swimming pool from drowning, a more common cause of death. In spite of this fact, solutions are actively sought to protect people from sharks; solutions that often involve the deaths of a much wider variety of marine life. Public surveys and focus groups conducted in Australia and South Africa reveal that, while beachgoers in Australia regarded safety as most important, beachgoers in South Africa thought that conservation was crucial. This need for conservation has led to a government-funded project to increase shark-spotting efforts as a preventative measure, rather than a range of more ecologically damaging alternatives (Dudley and Cliff, 2010). The Shark Spotting Programme created local employment, while protecting the great white sharks through an advanced-warning shark mitigation strategy.

Positive mitigation programmes have been funded by the governments of two Australian states (Western Australia and New South Wales). This funding has improved surveillance methods (through spotter aircraft, drones, watch towers, jetskis and sonar technologies) all focused on the early detection of sharks and eliciting rapid messaging to Surf Life Saving Clubs to initiate a retreat from the water by the public. The development of more user-friendly technologies, such as 'Sharksmart' tweet feeds, 'Sharktivity' and 'Global Shark Tracker' apps for mobile phones is making it possible for us to check where the sharks are as easily as we check our Facebook accounts before we head to the beach. Funded research projects enable scientists who are working on the complex sensory systems of sharks to understand which key environmental factors drive or provoke certain shark behaviours that might compromise human safety. By understanding how sharks perceive light, sound, odours, electromagnetic fields and temperature, marine neurobiologists hope to develop sensory deterrents to minimize the risk to both sharks and humans when they share the marine environment. Much of the cutting-edge work on sharks is being done at The University of Western Australia in the newly developed field of neuroecology (Collin, 2012), where research is focused on understanding how sharks perceive their three-dimensional environment, the relative importance of their many, well-developed, sensory modalities and how we can either over-stimulate these senses to change their behaviour and/or mask the biophysical cues that humans produce (and sharks can detect) that might mistakenly provoke an attack. Sharks are integral to the oceans and oceans start at the shoreline (Figure 1.16). Mitigation strategies may minimise the risk of having a negative encounter with a shark, but oceans are wild, untamed spaces, offering no guarantees of safety to those who enter them.

A combined suite of interdisciplinary strategies that incorporates marine biology, neurobiology, social science, cultural science and the creative arts is crucial to our goal of understanding and responding positively to the needs of sharks in our oceans. Drawing on recent research across these disciplines, *Sharks in the Arts* focuses chiefly on the ways in which sharks have been represented in society in order to demonstrate the enormous contribution that sharks have made to human culture. By increasing our understanding of the relationship between humans and sharks, it is hoped that we can alert the global community to the importance of sharks as part of their cultural heritage, thereby further promoting the need to respect and conserve them. Better strategies need to be devised for living harmoniously with these complex creatures. In the problem may lie the solution. Since sharks have been emotionally mapped negatively, it must be possible to remap them positively.

In recent years there have been signs of growing respect among some conservation groups for the role that sharks play. There has been a steady rise in the number of charities that have made sharks the focus of their efforts. Such organisations as Project Aware, Shark Friendly Marinas, The Shark Trust, Shark Savers, Bite-Back, Save Our Sharks, The Australian Marine Conservation Society, The Shark Research Institute, Shark Angels and Sea Shepherd actively promote the importance of sharks to the oceans through a wide range of strategies. Many of them raise funds to aid conservation efforts and have active websites that engage with the public internationally to bring them the latest news of sharks in peril. There are also now several useful single-authored works that focus on aspects of shark culture. Those that most responsibly and usefully treat sharks in the context of culture are Dean Crawford's *Shark* (London: Reaktion Books, 2008), an excellent framework for any study of sharks within culture; David Owen's *Shark: In Peril in the Sea* (Auckland, NZ: Allen & Unwin, 2010); Juliet Eilperin's *Demon Fish: Travels Through the Hidden World of Sharks* (New York: Pantheon, 2011), an engaging journalistic treatment of sharks through her personal experience of exploring the world of sharks and shark culture; and Colin Speedie's *A Sea Monster's Tale: In Search of the Basking Shark* (Plymouth: Wild Nature Press, 2017), a cultural history of the hunting of, and more recent conservation efforts surrounding, one species of shark.

Sharks in the Arts aims to offer authoritative and fresh perspectives on shark representations within more detailed and specific disciplinary contexts. It also engages with representations not generally available that self-consciously attach themselves to an agenda of positive representation with the aim of promoting conservation.

In order to overcome the fear of sharks, it is first necessary to recognize that this fear is a cultural construction, and then to understand that humans are responsible for the impact that culturally constructed fear has had on the lives of sharks and the marine environment more generally. There can be no doubt that provoked fear has historically led to a refusal to tolerate sharks in the oceans when humans choose to enter that space.

Organized culls, drumlining, shark nets and so on have too often been government responses to so-called public pressure to 'do something about the sharks'. Sharks have a key role to play in maintaining the health of the oceans, and interfering with it causes real damage to the ecosystem. In recognising the role that human emotion has played in the lives of sharks, it is important that attitudes are carefully managed so that more appropriate decisions can be made about the fate of sharks (Figure 1.17).

References

Primary works are derived from Chadwyck-Healey Literature Collections online.

Akcesme, B. (2013) An Ecocritical Approach to English Language Teaching. *Dil ve Edebiyat Egitimi Dergisi.* 2.8, 96–117.

Anonymous (1569) *The True Discription of this Marvellous Straunge Fishe.* London: Thomas Coldwell.

Atkinson, C.J.L. and Collin, S.P. (2012) Structure and Topographic Distribution of Oral Denticles in Elasmobranch Fishes. *Biological Bulletin*, 222.1, 26–34.

Atkinson, C.J.L., Martin, K.J., Fraser, G.J. and Collin, S.P. (2016) Morphology and Distribution of Taste Papillae and Oral Denticles in the developing Oropharangeal Cavity of the Bamboo Shark, *Chiloscyllium punctatum. Open Biology* 5: 1759–1769 doi:10.1242/bio.022327.

Ball, K. (2008) Primal Revenge and Other Anthropomorphic Projections for Literary History. New *Literary History.* Summer, 533–563.

Ballantyne, R.M. (1858) *The Coral Island.* London: T. Nelson & Sons.

Beaglehole, J.C. (ed.) (1961) *The Journals of Captain James Cook on his Voyages of Discovery 1772–1775.* Vol II. Cambridge: Hakluyt Society, CUP.

Beaglehole, J.C. (ed.) (1962) *The Endeavour Journal of Joseph Banks 1768–1771.* Sydney: Halstead Press.

Beaglehole, J.C. (ed.) (1967a) *The Journals of Captain James Cook on his Voyages of Discovery 1776–1780.* Vol III Part I. Cambridge: Hakluyt Society, CUP.

Beaglehole, J.C. (ed.) (1967b) *The Journals of Captain James Cook on his Voyages of Discovery 1776–1780.* Vol III Part II. Cambridge: Hakluyt Society, CUP.

Bedoyere, C. (2013) *Discover Sharks: Great White Sharks.* London: QED Publishing.

Behrendt, K. A Special Way of Being Afraid (2010) *Philosophical Psychology*, 23.5, October, 669–682.

Bensley, N., Woodhams, J., Pattterson, H., Rodgers, M., McLoughlin, K., Stobutzki, I. and Begg, G. (2010) 2009 Shark Assessment Report for the Australian National Plan of Action for the Conservation and Management of Sharks. Canberra: Fisheries and Forestry Department of Agriculture, Bureau of Rural Sciences.

Borgi, M. and Cirulli, F. (2015) Attitudes toward Animals Among Kindergarten Children: Species Preferences, *Anthrozoos*. 28, 45–59.

Brayton, D. (2012) *Shakespeare's Ocean.* Virginia: University of Virginia Press.

Brosch, T., Pourtois, G. and Sander, D. (2010) The Perception and Categorisation of Emotional Stimuli: A Review. *Cognition and Emotion*. 24.3, 377–400.

Brathwait, R. (1658) *The Captive-Captain.* Chadwyck-Healey English Poetry Database.

Brown, J. (2007) Overfishing Large Sharks Impacts Entire Marine Ecosystem, Shrinks Shellfish Supply. *Science Daily*, 29 March. www.sciencedaily.com/releases/2007/03/070329145922.htm.

Bruce, B. (2015) A Review of Cage-Diving Impacts on White Shark Behaviour and Recommendations for Research and the Industry's Management in New Zealand. Hobart: CSIRO Marine & Atmospheric Research for the Department of Conservation, New Zealand.

Cantor, J.R. (2009) Fright Reactions to Mass Media. In Bryant J., Oliver M.B. (eds), *Media effects: Advances in Theory and Research* (3rd ed., pp. 287–303). New York, NY: Routledge.

Carey, B. (2005) The Science of Shark Attacks and How to Avoid Them. http://www.livescience.com/3892-science-shark-attacks-avoid.html (Accessed 30.05.2017).

Cartoonstock. http://www.cartoonstock.com/directory/s/shark.asp. Accessed 13.05.2012.

Casey, S. *Devil's Teeth (*2006) Owl Books: New York.

Cenki, P. (2006) *This Vast Book of Nature.* Iowa: University of Iowa Press.

Cerullo, M.M. (2005) *The Truth About Great White Sharks.* Minneapolis: Lerner Publishing Group.

Chapman, B. B. and McPhee, D. (2016) Global Shark Attack Hotspots: Identifying Underlying Factors Behind Increased Unprovoked Shark Bite Incidence. *Ocean and Coastal Management.* 133, 72–84.

Collin, S. P. and Whitehead, D. (2004) The Functional Roles of Passive Electroreception in Non-Electric Fishes. *Animal Biology.* 54, 1–25.

Collin, S. P. (2012) The Neuroecology of Cartilaginous Fishes: Sensory Strategies for Survival. *Brain Behavior and Evolution.* 80.2, 80–96.

Collin, S. P. and Hart, N. S. (2015) Vision and Photoentrainment In Fishes: The Effects of Natural and Anthropogenic Perturbation. *Integrative Zoology.* 10, 15–28. doi: 10.1111/1749-4877.12093.

Collin, S. P., Kempster, R. and Yopak, K. E. (2015) How Elasmobranchs Sense their Environment. In: Physiology of Elasmobranch Fishes: Structure and Interaction with Environment (Shadwick, R. E., Farrell, A. P. and Brauner C. J. eds). Vol 34A, 1st Edition, Elsevier, New York, 19–99.

Copley, J. S. (1778) *Watson and the Shark.* MFA, Boston.

Cotton, C. (1667) *Scarronides.* Chadwyck-Healey English Poetry Database.

Crawford, D. (2008) *Shark.* London: Reaktion Books.

Crossley, R., Collins, C. M., Sutton, S. G. and Huveneers, C. (2014) Public Perception and Understanding of Shark Attack Mitigation Measures in Australia. *Human Dimensions of Wildlife.* 19.2, 154–165.

Curtis, T., Bruce, B., Cliff, G., Dudley, S., Klimley, A. P., Kock A., Lea, R.N., Lowe, C., McCosker, J. E., Skomal, G., Werry, J. M. and West, J. G. (2012) Responding to the Risk of White Shark Attack: Updated statistics, prevention, control methods and recommendations. In: *Global Perspectives on the Biology and Life History of the White Shark*, M. Domeier (ed.) Florida, CRC Press. 477–509.

Cumberland, R (1789) *The Imposters.* Chadwyck-Healey English Drama Database.

Dallas, R. C. (1823) *Amabel.* Chadwyck-Healey English Poetry Database.

Dawes, B. (2014) The Oceanic Imagination: Canadian and Australian Contributions to a Trans-Indigenous Methodology'. *The Canadian Journal of Native Studies,* 34.2, 65–84.

Dampier, W. (1931) *Voyages and Discoveries* with an introduction and notes by Clennell Wilkinson. London: The Argonaut Press.

Dell'Apa, A., Chad Smith, M. and Kaneshiro-Pineiro, M.Y. (2014) The Influence of Culture on the International Management of Shark Finning. *Environmental Management,* 54, 151–161.

De Raad, B. and Kokkonen, M. (2000) Traits and Emotions: A Review of their Structure and Management. *European Journal of Personality*, 14, 477–496.

Diaper, W. 1712. *Nereides: or Sea-Eclogues.* Chadwyck-Healey English Poetry Database.

Dudgeon, C. L., Blower, D.C., Broderick, D., Giles, J. L., Holmes, B. J., Kashiwagi, T., Kruck, N. C., Morgan, J. A. T., Tillett, B. J. and Ovenden, J. R. (2012) A Review of the Application of Molecular Genetics for Fisheries Management and Conservation of Sharks and Rays. *Journal of Fish Biology.* 80.5, 1789–843.

Dudley, S. F. J. and Cliff, G. (2010) Shark Control. In Carrier, J. C., Musick, J. A. and Heithaus, M. R. (eds) *Sharks and Their Relatives II Biodiversity, Adaptive Physiology, and Conservation.* Florida, CRC Press; Taylor and Francis, 567–591.

Eilperin, J. (2011) *Demon Fish: travels through the hidden world of sharks.* New York: Pantheon.

Evans S. (2014) Shark Week in an Era of Infotainment. Paper presented at the NEXUS Interdisciplinary Conference, Knoxville, TN.

Fallows, C., Gallagher, A.J. and Hammerschlag, N. (2013) White Sharks (*Carcharodon carcharias*) Scavenging on Whales and Its Potential Role in Further Shaping the Ecology of an Apex Predator. *PLOS ONE,* 0060797 (Accessed 19 May 2013).

Ferretti, F., Worm, B., Britten, G.L., Heithaus, M.R. and Lotze, H. K (2010) Patterns and Ecosystem Consequences of Shark Declines in the Ocean. *Ecology Letters*, 13, 1055–1071.

Field, I.C., Meekan, M.G., Buckworth, R.C., Bradshaw, C.J. A (2009) Susceptibility of Sharks, Rays and Chimaeras to Global Extinction. In: Sims, D.W. (ed.) *Advances in Marine Biology*. 56, Cam. Ma., Elsevier, 275–363.

Finding Nemo (2003) Pixar Animation Studios, Walt Disney Pictures.

France-Porcher, I. (2017) *The True Nature of Sharks*. California: Independent.

Gilbert, P.W. (1970) Studies on the anatomy, physiology, and behavior of sharks. *Mote Technical Report No. A-1970b.* Mote Marine Laboratory, Sarasota, FL.

Gosselin, P. and Simard, J. (1999) Children's Knowledge of Facial Expressions of Emotions: Distinguishing Fear and surprise. *Journal of Genetic Psychology.* 160. 2, 181–193.

Griffin, E., Miller, K.L., Freitas, B. and Hirschfield, M. 2008. Predators as prey: Why Healthy Oceans Need Sharks. Washington: Oceana. www.oceana.org.

Guerin, F. (2015) *On Not Looking: The Paradox of Contemporary Visual Culture.* Abingdon and New York: Routledge.

Gyurak, A., Gross, J.J. and Etkin, A. (2011) Explicit and Implicit Emotion Regulation: A Dual-Process Framework. *Cognition and Emotion*, 25.3, 400–412.

Hammerschlag, N., Martin, R.A., Fallows, C., Collier, R.S. and Lawrence, R. (2012) Investigatory Behaviour toward Surface Objects and Nonconsumptive Strikes on Seabirds by White Sharks, *Carcharodon carcharias*, at Seal Island, South Africa (1997–2010). In *Global Perspectives on the Biology and Life History of the White Shark.* Domeier, M.L. (ed.) Florida: CRC Press, 91–104.

Hart, N.S., Theiss, S.M., Harahush, B.K. and Collin, S.P. (2011) Microspectrophotometric Evidence for Cone Monochromacy in Sharks. *Naturwissenschaften,* 98: 193–201.

Harvey, E.D. (2013) Beastly Physic. *Shakespeare Studies.* 41, 114–124.

Haynes, M. (2013) Regulating Abjection: Disgust, Tolerance, and the Politics of The Cove. *English Studies in Canada.* 39.1, March, 27–50.

Heupel, M.R. and Simpfendorfer, C.A. (2010) Science or Slaughter: Need for Lethal Sampling of Sharks. *Conservation Biology*. 24.5, 1212–1218.

Hirst. D. (1991) http://en.wikipedia.org/wiki/The_Physical_Impossibility_of_Death_in_the_Mind_of_Someone_Living (Accessed 13.05.2012).

Hood, T. n.d. *Sally Simpkin's Lament.* Chadwyck-Healey English Poetry Database.

Howard, R.A. (1984) On Fates Comparable to Death. *Management Science*. 30.4, 407–422.

International Shark Attack File: https://www.floridamuseum.ufl.edu/fish/isaf/home/

International Union for the Conservation of Nature: https://www.iucn.org

Janis I.L., Feshbach S. (1953) Effects of Fear-Arousing Communications. *Journal of Abnormal and Social Psychology*. 48, 78–92. doi:10.1037/h0060732.

Jaws (1975) Zanuck/Brown Productions, Universal Pictures.

Kalmijn A.J. (1971) The electric sense of sharks and rays. *Journal of Experimental Biology.* 55, 371–383. PMID: 5114029.

Kempster, R.M., Hart, N.S. and Collin, S.P. (2013) Survival of the Stillest: Predator Avoidance in Shark Embryos. *PLoS ONE* 8.1: e52551. doi:10.1371/journal.pone.0052551.

Kempster, R.M., Egeberg, C.A., Hart, N.S. and Collin, S.P. (2015) Electrosensory-Driven Feeding Behavior of the Port Jackson Shark (*Heterodontus portusjacksoni*) and

Western Shovelnose Ray (*Aptychotrema vincentiana*). *Marine and Freshwater Research.* doi.org/10.1071/MF14245.

Kempster, R.M., Egeberg, C.A., Hart, N.S., Ryan, L., Chapuis, L., Kerr, C., Schmidt, C., Huveneers, C., Gennari, E., Yopak, K.E., Meeuwig, J., and Collin, S.P. (2016) How close is Too Close? The Effect of Non-Lethal Electric Fields on White Shark Behaviour. *PLoS ONE* 11(7), e0157717. doi:10.1371/journal.pone.0157717.

Kerridge, R. (2013) Ecocriticism. *Year's Work in Critical & Cultural Theory.* Jan, 21.1, 345–374.

Kidd, A.H. and Kidd, R.M. (1996) Developmental Factors Leading to Positive Attitudes Toward Wildlife and Conservation. *Applied Animal Behaviour Science.* 47.1–2, 119–125.

Kock, A., Titley, S., Peterson, W., Sikweyiya, M., Tsotsobe, S., Colenbrander, D., Gold, H. and Oelofse, G. (2010) Shark Spotters: A Pioneering Shark Safety Program in Cape Town, South Africa. In: Domeier, M.L. (ed.) *Global Perspectives on the Biology and Life History of the White Shark.* Florida, CRC Press; Taylor & Francis. 447–466.

Lanning, M. (2000) *African Americans in the Revolutionary War.* New York: Kensington. Cited in Wickipedia. http://en.wikipedia.org/wiki/African_Americans_in_the_Revolutionary_War (Accessed 13.05.2012).

Lisney, T.J., Theiss, S.M., Collin, S.P. and Hart, N.S. (2012) Vision in Elasmobranchs and their Relatives: 21st Century Advances. *Journal of Fish Biology.* 80, 2024–2054.

McCalman, I. and Frawley, J. (eds) (2014) *Rethinking Invasion Ecologies from the Environmental Humanities.* Abingdon & New York: Routledge.

McFadden, E.B., Johnson, C.S. (1978) Color and reflectivity of sea-survival equipment as related to shark attack. Report FAA-AM-78-1. In: McFadden, E.B. (ed.), *Flotation and Survival Equipment Studies.* FAA Civil Aeromedical Institute, Oklahoma City, OK, 41–5.

Muter, B.A., Gore, M.L., Gledhill, K.S., Lamont, C. and Huveneers, C. (2013) Australian and US News Media Portrayal of Sharks and their Conservation. *Conservation Biology.* 27.1, 187–196.

Myers, R.A., Baum, J.K., Shepherd, T.D., Powers, S.P. and Peterson, C.H. (2007) Cascading Effects of the Loss of Apex Predatory Sharks from a Coastal Ocean. *Science.* 315, 1846–1850.

Myrick, J.G. and Evans, S.D. (2014) Do PSA's Take a Bite Out of *Shark Week*? The Effects of Juxtaposing Environmental Messages with Violent Images of Shark Attacks. *Science Communication.* 36.5, 544–569.

Neff, C. and Hueter, R. (2013) Science, Policy, and the Public Discourse of Shark 'attack': a Proposal for Reclassifying Human-Shark Interactions. *Journal of Environmental Studies and Science.* 3, 65–73. DOI 10.1007/s13412-013-0107-2.

Nelson, D.R. (1983) Shark Attack and Repellency Research: an Overview. In: Zahuranec, B.J., ed. *Shark Repellents from the Sea. New Perspectives. AAAS Selected Symposium 83.* Westview Press, Inc., Boulder, CO, 11–74.

Newman, A., Robinson, P.R. 2005. Cone Visual pigments of Aquatic Mammals. *Visual Neuroscience* 22(6), 873–9.;

O'Keeffe. J. (1793) *The world in a village.* Chadwyck-Healey English Poetry Database.

Owen, D. (2010) *Shark: in Peril in the Sea.* Auckland, NZ: Allen & Unwin.

OED Online. March 2017. Oxford University Press. http://www.oed.com.ezproxy.library.uwa.edu.au/view/Entry/177548?rskey=8tUmT1&result=1 (Accessed June 08, 2017).

Peichl L., Behrmann G., Kröger R.H.H. (2001) For Whales and Seals the Ocean is not Blue: a Visual Pigment loss in Marine Mammals. *European Journal of Neuroscience* 13, 1520–8.

Peron, F. and De Freycinet. L. (2006) *Voyage of Discovery to the Southern Lands by Francois Peron Continued by Louis De Freycinet. Second Edition 1824, Books I–III Comprising Chapters I–XXI.* Adelaide: The Friends of the State Library of South Australia.

Rasch, L. J., Martin, K. J., Cooper, R. L., Metscher, B. D., Underwood, C. J. and Fraser, G. J. (2016) An Ancient Dental Gene Set Governs Development and Continuous Regeneration of Teeth in Sharks. *Developmental Biology.* 2, 347–370.

Russell, G. (2015) Mick Fanning Shark Attack: Internet Adds Punchy Humour to Surfer's Tale. *The Guardian.* https://www.theguardian.com/sport/2015/jul/20/mick-fanning-shark-attack-internet-adds-punchy-humour-to-surfers-tale (Accessed 30.05.2017).

Saldanha, A. (2016) A New Earth: Deleuze and Guattari in the Anthropocene. *Deleuze Studies.*10.4, 427–439.

SAMS (2013). *Scientifically designed shark deterrent wetsuits and watersport products.* [Cited 28 Oct 2013.] Available from: http://www.sharkmitigation.com/

Shakespeare, W. *Macbeth (*1606; 2015) Clark, S. and Mason, P. (ed.). London: Bloomsbury.

*Sharknado (*2013) Southward Films, Syfy.

Sharkwater (2006) SW Productions, Diatribe Pictures, Sharkwater Productions.

Shark Tank (2009–) http://abc.go.com/shows/shark-tank.

Shark Week (1988–) http://www.adweek.com/adfreak/5-great-shark-ads-honor-shark-week-133886 (Accessed 13.05.2012).

Shoba, K. N. (2012) From Nature Writing to Ecocriticism – An Evolutionary Outline of Ecocritical Writing. *Language in India*, Aug, 12.8, 441–453.

Slovic, P., Finucane, M. L., Peters, E., and MacGregor, D. G. (2004) Risk as Analysis and Risk as Feelings: Some Thoughts about Affect, Reason, Risk, and Rationality. *Risk Analysis* 24.2, 311–322.

Stevens. G. A. (1767) *The French Flogged.* Chadwyck-Healey English Drama Database.

Swaim, K. (1973) Lycidas and the Dolphins of Apollo. *The Journal of English and Germanic Philology*, 72.3, 340–349.

Taronga.org. au. https://taronga.org.au/animals-conservation/conservation-science/australian-shark-attack-file/latest-figures.

Techera, E. J. and Klein, N. (2014) *Sharks: Conservation, Governance and Management.* Abingdon and New York: Routledge.

Tester, A. L., Nelson, G. J., Daniels, C. I. (1968) Test of NUWC Shark Attack Deterrent Device. *NUWC TP* 53. Naval Undersea Warfare Centre, Pasadena, CA.

This is Your Ocean: Sharks (2012) Status Productions.

Toomey, J. *Sherman's Lagoon* (1991–) http://shermanslagoon.com/meet-the-cast (Accessed 19.05.2017)

Toy Story (1995) Pixar Animation Studios, Walt Disney Pictures.

Weisz, A. (2011) Wolfgang Leander. *X-Ray Magazine.* Oct. 13, 23–26.

Werry, J. M., Planes S., Berumen, M. L., Lee, K. A., Braun, C. D. and Clua, E (2014) Reef-Fidelity and Migration of Tiger Sharks, *Galeocerdo Cuvier*, Across the Coral Sea. *PLOS ONE*, https://doi.org/10.1371/journal.pone.0083249.

Wilkinson, C. (ed.) (1929) *William Dampier.* London: John Lane and Bodley Head Ltd.

Witte, K. (1992) Putting the Fear Back into Fear Appeals: The Extended Parallel Process Model. *Communication Monographs.* 12, 329–349.

Wolfer, D. (2016) *The Shark Caller.* Melbourne: Random House, Australia.

Worm, B., Davis, B., Kettemer, L., Ward-Paige, C. A., Chapman, D., Heithaus, M. R., Kessel, S. T. and Gruber, S. H. (2013) Global Catches, Exploitation Rates, and Rebuilding Options for Sharks. *Marine Policy.* 40, 194–204.

Yopak, K.E., Lisney, T.J., Darlington, R., Collin, S.P., Montgomery, J. and Finlay, B. (2010) Brain Scaling from Sharks to Primates: a Highly Conserved Vertebrate Pattern. *PNAS*. 107, 12946–12951.

Yopak, K.E., Lisney, T.J. and Collin, S.P. (2014) Not all Sharks are 'Swimming Noses': Variation in Olfactory Bulb Size in Cartilaginous Fishes. *Brain Structure and Function*. DOI:10.1007/s00429-014-0705-0.

Major shark conservation sites

http://www.projectaware.org
http://www.saveoursharks.com.au
http://sharkfriendlymarinas.org
http://sharkangels.org
http://www.sharksavers.org/en/home/
http://www.seashepherd.org
http://www.sharks.org
http://www.sharktrust.co.uk
http://www.bite-back.com
https://www.marineconservation.org.

2 Establishing the reputation of sharks

The origin of the word 'shark' is obscure. It seems that there are at least three possible contenders for the etymology of 'shark': one French, one German and one Mayan. The *OED* cites Walter Skeat's explanation that the French verb 'cherquier', to seek, may have given its name to the two nouns 'shirk' and 'shark', the parasite and the fish, even as the dictionary contests this etymological trajectory. 'Sherk' becoming 'shirk' would seem to support the idea of a German origin. The scholar Tom Jones makes a case for 'Xoc' being 'the only Mayan loan word in the English language'. If, as he suggests, the pilot Bartolome Gonzales, a native of Campeche, did indeed teach the crew of John Hawkins' *Minion* his language, it is odd that only this word stayed in the minds of the crew (Jones, accessed 2012). The Cornish word 'skarkas' might also provide a root word in a very muddled origin story. We cannot ignore the possibility that they all have their place in the English language from each of these sources, and that these 'sharks' which were being used in different contexts through popular culture fused together, creating etymological confusion.

Sharks have, of course, been present in the oceans of the world throughout human history, and for roughly 450 million years before any human walked the planet. They interacted with many cultures long before the early modern period, playing their part in the cultural development of civilizations (Figure 2.1). A proper exploration of that influence, however, requires clarity over the nuances of foreign, often obsolete, words in historic settings. The same word might, for example, be translated as shark, or toothed fish, or dogfish, or big fish, or sea monster, or whale, or indeed some other specific species. Here, writ large, is the old problem of how to render 'dahg' from the book of Jonah or 'κήτους' from St Matthew's Gospel. In the same vein, are Oppian's 'ketea' in his *Halieutica*, insatiably hungry fish with two powerful jaws, identifiable as sharks? That might be a reasonable assumption, but it is still an assumption. Presumably Aristotle's cartilaginous 'selache' are sharks? When Herodotus refers to fish devouring Persian seamen in the Aegean, or when Pliny the Elder writes of fish attacking pearl divers, do they refer to sharks of one species or another? We cannot now say so for sure. All that can safely be suggested is that primitive myths travelled a separate path to the more scientific tradition through to the Middle Ages, when the two gradually became confused and blended. The great serpent can be terrestrial or marine.

Shark in print

In keeping with these obscure and contradictory roots, the first time we encounter the word 'shark' in print, we are almost certainly meeting a misprint. 'Sharke' replaces 'sharpe' in a book on surveying and improvements to land written by John Fitzherbert, and published in 1523. It is only with the landing of a strange fish caught by mackerel fishermen in the English Channel in June 1569, that the shark is named, described and depicted in an opportunist leaflet announcing its display at the *Red Lion* in Fleet Street. Seventeen feet long and three feet broad, the splendid fish had three rows of teeth in each jaw, 'marvaylous sharpe and very short' (C. R., 1569). Typically, in an Elizabethan England avid for novelty, some Londoners ate the flesh and found it not wholly displeasing. The shark's heart, authors noted, was as big as that of an ox.

Over the next twenty years or so the word 'shark' seems to become familiar throughout literate English society, without leaving traces in the surviving printed record. This high Elizabethan 'dark age' is tantalising, because by the time the physician, author and traveller Thomas Lodge published *Euphues Shadow* in 1592, 'shark' needed no gloss, no laboured explanation such as that supplied in 1569. Indeed, Lodge's informed London readership was assumed to have some familiarity with the creature's behaviour. In an early reference to a bond that intrigued so many writers in centuries to come, Lodge referred to the relationship between shark and pilot fish, as inseparable companions (Figure 2.2). This familiarity may owe something to travellers' tales, the yarns of Hawkins' seamen and their Elizabethan successors on the deep ocean which Lodge heard at first hand, eventually finding its way into the ballads, poetry and drama of the age. Richard Hakluyt, the paramount contemporary collector of such narratives, noted that the Spanish named the western tip of the island of Hispaniola Cape Tiburon, likening this promontory to the 'snout of a Tiburon, or sharke-fish', while also recording in passing, as in Luke Ward's account of his voyage to China aboard the *Edward Bonaventure* during 1582, the pastime of fishing for sharks (Hakluyt, 1626). By this time readers are assumed to be able to picture everything necessary, without gloss.

The courtier and historian Sir Walter Raleigh, upholding the plausibility of the Ark story from the Book of Genesis, and indeed allowing for a form of evolution in his highly popular *History of the World*, argues that 'The dog-fish of England is the Sharke of the South Ocean: For if colour or magnitude made a difference of Species, then were the *Negro's,* which wee call the blacke-mores, *non animalia rationalia,* not men, but some kinde of strange beastes' (Ralegh, 1614: 1, 7, ix).

In one of the best early descriptions, Richard Hawkins also likens the exotic new creature to his more prosaic cousin:

> The shark is a Fish like unto those which wee call Dogge-fishes, but that he is farre greater. I have seene of them eight or nine foote long; his head is flatt

> and broad, and his mouth in the middle, underneath, as that of the Seate; and he cannot byte of the bayte before him, but by making a halfe turne; and then he helpeth himselfe with his tayle, which serveth him in stead of a Rudder. His skinne is rough (like to the Fish which we call, a rough Hound) and russet, with reddish spottes, saving that under the belly he is all white.
>
> (Hawkins, 1622: 43–4)

The rows of teeth are described as being 'as sharpe as nailes'. These stereotypical shark characteristics will be listed time and time over hereafter. Again, Hawkins refers to the attendant 'pilats-fishes'.

Raleigh's reflections show how contemporaries gradually appreciated that sharks swam the oceans in more than one variety. Hakluyt's successor as chronicler of English navigational enterprise, Samuel Purchas, provides by way of illustration some information on the South American littoral: 'Upon this Coast are many sorts of Sharkes, and in it are found sixe or seuen kinds of them; it is a very fierce and cruell fish, and kils many persons, especially those that swimme' (Purchas, *Pilgrimes*, 1625: 1314). Describing the marine life off Sierra Leone in 1607, William Finch also sees diversity, even a profusion of species, in these tropical waters: 'fishes they have many of ordinary kind, and some which seemed to us extraordinarie: Mullets, Raies, Thorne-backes, Old-wiues with beetle Browes, Fishes like to Pikes, Gar-fishes, Caudllios (like Mackerell) Sword-fishes with snowts a yard long, toothed on both sides like the teeth of a Saw, Sharkes, Dog-fishes; Shark[e?]rs, like a Sharke, but having a snowt flat and broad like a Shovell' (Purchas, 1625: 416).

Something else of great importance occurred during the Elizabethan 'dark age' of the shark. Many perceived as well as actual characteristics of the fish were almost instantaneously applied to people. Like other tokens of the emergent English overseas Empire, words such as 'cannibal' and 'pirate', the 'shark' enjoys a period of fashion and topicality, when a literate, sophisticated man or woman enjoys dropping it into conversations, letters and pamphlets (Working, 2015).

By the 1620s, the human shark has clearly become a stock figure in contemporary writing. Characteristics developed early, and then stopped developing. The shark is a modernised member of a timeless picaresque crew, the sort of petty crook noted in the comedies of Aristophanes, Plautus or Terence:

> The Jealous Lovers, Pander, Gull, and Whore,
> The doting Father, Shark, and many more
>
> (Randolph, 1632)

In the seamy underworld of Jacobean London, neediness was for many an everyday predicament. But any understanding of the word differs in nuance rather than substance, allowing 'shark' to encompass many forms of opportunistic criminal, low-life behaviour. Some things were certain. The shark was to be avoided by honest people. The shark was 'cunning', a trickster, a common thief with attitude, as caught in this passage from the poet Robert Herrick's hugely popular *Hesperides* (1648):

> Shark when he goes to any publick feast,
> Eates to ones thinking, of all there, the least.
> What saves the master of the House thereby?
> When if the servants search, they may descry
> In his wide Codpeece, (dinner being done)
> Two Napkins cram'd up, and a silver Spoone
>
> (136)

Often he is too clever for his own good. The 'shark' is 'god-lesse' (Brathwaite, 1615). He is a perjurer (Dobson, 1644: 5). Like the eponymous fish, the human shark must predate: 'the best prey I ever shark'd for' (S. S., 1616: I1v). 'Italian, Spanish, English, Dutch, or French; Sharke what you will, you shall not sharke me out of my stagges leather gloves with Charlatan' (Belchier, 1618: D1v). 'A very Sharke, hee set me i'the nicke t'other night at *Primero*' (Jonson, 1620: IV: 3). A 'base, defamed' creature, for John Taylor he is the antithesis of another, older stereotype, the honest English archer (Taylor, 1621: C2). In a nod to the shark's speed and agility, swimming in his own drier element, the 'land-shark' – an early usage of a compound noun subsequently common – is slippery, and ready to put distance between himself and the scene of his crime (Taylor, 1622: D2v).

The flexibility of the word in the early seventeenth century encouraged use. Sharks could be found anywhere: in towns, in trades, in households, in the Army. By 1625, the universality of usage is demonstrated by an Edinburgh tract promoting settlement in the new world (Gordon, 1625: C). During the English Civil War rapacious, or desperate, troops 'shark for' provisions: perhaps plunder, perhaps a reference to the authorised sleight of hand that involved the issue of worthless promissory notes (*Mercurius cambro-Britannus*, 13 Jan. 1644:A4). Both generic and hostile, the word was one more useful epithet to hurl at your opponents in a pamphlet war.

A shark was, thus, the false friend, the discredited acquaintance, encountered almost anywhere in a fallen world. The 'Old Shark' in an epigram by the poet Thomas Bancroft stands in the realm of social acquaintance rather than business:

> *Sharke* bad me to his roast, but in the end
> Forc'd me to pay both for my selfe and *friend:*
> Thus (though a Coward) shew'd he mettle yet,
> In beating of me with a silver *spit*
>
> (Bancroft, 1639: Epigram 180)

There was no honour in the early modern shark. He was, consistently, a more despised, contemptible creature than his modern equivalent, lacking the business-shark's dark glamour. Humphrey Mill, scorning all forms of 'darkness', placed the 'shark' in base company: 'why should I favor crave, of thee that art a theefe, a sharke, a slave?' (Mill, 1639: no. 58).

Some of these excerpts demonstrate that 'shark' is by now a verb too. As Will Summers says in Samuel Rowley's *When You See Me, You Know Me*, of 1605: 'shark with me!' i.e. witness my influence over petty officialdom. Come and be unscrupulous, controlling, ready to try any means to an end.

Thomas Middleton used the word in a similar way in *A Mad World, My Masters* (1608): 'neither Pilfer, Cheat, nor Sharke,' says Sir Bounteous Progress in Act V. There doesn't seem much to choose between the three. Like the noun, the verb is soon everywhere. By the 1630s, indeed, it probably accounts for most surviving usages.

Occasionally an author refers to some specific characteristic of the oceanic shark, to emphasise the deficiencies of its human counterpart. In his exploration of London low-life, a popular theme since the days of Robert Greene and Thomas Nashe, Thomas Dekker likened the disappearance after defrauding his prey of a cozener, a variant of the common criminal colloquially known as the 'Jack in a Box', to the swimming and diving skills of the shark (Dekker, 1609: sub. 'Jack in a box').

All this while, the fish swims out at sea, mysterious, terrifying, poorly understood. Knowledge of the watery shark develops more slowly than the elaborate meanings attached to its human counterpart. In John Nicholl's true account of shipwreck, privation and survival, published in 1607, the prototype of so many factual and fictional tales, sailors navigate a leaky, makeshift boat to the Caribbean mainland, avoiding alligators and sharks with their 'three ranks of teeth set like a saw', ready to 'bite off a mans thigh at one snatch'. Starvation and other privations among the cannibal islands kill most of Nicholl's party. But God does at least preserve them from the shark. Purchas's collection of travellers' tales included stories of shark attacks, but the threat did not only work one way (Purchas, 1625:1209). Richard Hawkins gave a vivid picture of the animosity already harboured by sailors towards this 'unlucky' and apparently malevolent fish:

> Every day my Company tooke more or lesse of them, not for that they did eate of them (for they are not held wholesome; although the Spaniards, as I haue seene, doe eate them) but to recreate themselues, and in revenge of the iniuries received by them; for they liue long, and suffer much after they bee taken, before they dye. At the tayle of one they tyed a great logge of wood, at another, an emptie *Batizia* well stopped; one they yoaked like a Hogge; from another, they plucked out his eyes, and so threw them into the sea.
>
> (Hawkins, 1622:43–4)

Alongside a characteristic denigration of foreigners' eating habits, the potent early modern importance attached to 'revenge' is brought to bear here against an animal rather than a human opponent. There is something of Melville's 'Shark massacre' in *Moby-Dick*, something of the Patrick O'Brian books here. The sailors' revenge was 'for that [the shark] deprived them of swimming, and fed on their flesh being dead' (Hawkins, 1622:43–4). But implicitly, fear is at work here too. People were willing 'to believe the fantastic worst' (Crawford, 2008:97). The shark is branded a man-eater with no evidence other than, one supposes, reputation. Off Madagascar and Mozambique, in the 1620s, the well-born Welsh seaman and author Thomas Herbert's crew caught 'a Sharke, a man-eating Monster'. The fish 'was nine Foot long and a halfe by rule, in her paunch were fiftie fiue yong Sharkes euery one aboue a Geometricall foot in length' (Herbert, 1634:21). Herbert brought forward no evidence to prove that this 'man-eater' ever ate men. In an early description of sharks off New England, William Wood dwelt on another element in the evolving stereotype: the power of a shark's bite. Hooks and lines had to be adapted, if you were to catch them, but he conceded that once caught the shark was of limited nutritional or commercial value (Wood, 1634).

It was commonly accepted that sharks stuck to the deep sea, and did not infest rivers and inshore waters. As Pliny contended (incorrectly) when writing of his 'dogfish' centuries earlier, it was a fish that 'never commeth neare the shore, but is constantly in the depth of the Ocean', like 'the Flaile-fish, and others of that noxious kinde' (Anon., 1642:2). In contrast to its human counterpart, and for whatever reason, it kept its distance.

Within the intense battle for attention, reception and preferment fought out by religious controversialists and the writers of sermons, it was important to speak in the idioms of the day. Vogue words that rolled with the tongue always played well. So doctrinal opponents or insincere backsliders become, among other things, sharks intent on damaging and predating the true church. What, asked the Anglican Humphrey 'silver tongue' Sydenham rhetorically, are Catholic ceremonies worth? What is 'a 500 yeares indulgence, but the sharke and legerdemaine of your Lord God the Pope?' (Sydenham, 1626:22). Robert Parker, in a discourse against the sign of the cross in ceremony, suggested that such gestures were futile: 'it is so thinne a sharke, that our own Church cannot countenance it neither, though commaund it most vnworthy'. A thin shark is a poor creature, almost deserving of pity, but certainly meriting no respect (Parker, 1607: sect. 11). Robert Southey's 'Shakespeare of the Puritans', the preacher Thomas Adams, in his early work *Heaven and earth reconciled* (1612), scorned fellow clerics who entered the Church only when other career options failed, and who then 'sharke[d] out a living' once ordained. 'Somtimes', he insisted, using the early modern vogue for inversion, 'such Beggars are made Priestes, when good Priestes are made Beggars.' The lack of discrimination in the shark-fish's rapacious feeding also found a dismissive echo in the way that some divines 'sharke about' in the Scriptures, looking for suitable sermon material (Goodwin, 1642:123). This combination of rapacity and dishonesty was picked up by other writers commenting on a worldly Church, Leighton and Milton among them (Edwards, 2009:106–8). It offered so many easy targets.

When a shark is encountered in English verse for the first time, he is human, and blessed with the by now accepted attributes. The London pamphleteer Samuel Rowlands' poem 'A Sharke', published in his 1609 collection *The knave of clubbs*, recorded the contrasting fortunes of two London 'sharks' intent on cheating their way to a free tavern meal. One comes up with a scam which pays off; but the other, copying his colleague too closely, is seen for what he is and beaten up by the indignant victims. The moral of the story is, of course, to be original in your schemes. Another member of the same family appeared in the 'water poet' John Taylor's *The sculler rowing from Tiber to Thames* (1612), although in this Epigram the central figure is sated and dissipated, rather than actively mischievous:

> Great Captaine *Sharke* doth wonderfully muse,
> How he shall spend the day that next ensues:
> Theres no Play to be plaid but he hath seene,
> At all the Theaters he oft hath been:
> And seene the rise of Clownes, and fall of Kings,
> Which to his humor no contentment brings.
> And for he scornes to see a Play past twice:
> Heele spend the time with his sweet Cockatrice
>
> (Taylor, 1612: Epigram 22)

This corruptible, humorous, and yet slightly elusive human figure held wide appeal for the popular poet.

The aquatic shark was not entirely neglected. In a posthumously published 1613 verse evocation of the pleasures and skills involved in freshwater fishing, John Dennys of Gloucestershire specifically excluded from his work the fish and other creatures of the open ocean. Perhaps unsure of his facts when dealing with salt water, he named the 'tuberone' and the 'sharke' separately, the first attended (neutrally) by its 'guard' – the pilot fish – and the latter ravenous and scavenging:

> Nor with that Fish that beareth in his snout
> A ragged sword, his foes to spoile and kill;
> Nor that fierce *Thrasher,* that doth fling about
> His nimble flayle, and handles him at will:
> The rauenous *Sharke* that with the sweepings out
> And filth of ships doth oft his belly fill.
> The *Albacore* that followeth night and day
> The flying Fish, and takes them for his pray.

Sharks may be treated with disgust, and fear, by English sailors, authors and readers, but early writings also appreciated that the animal might be venerated by 'ignorant' indigenous peoples. Occasional references in early literature highlight a point made by Dean Crawford: 'the cultures that have lived most closely with sharks have traditionally viewed them not as monsters but as religious entities' (Crawford, 2008: 47). Among the Inca in Peru, for example, 'The Fishers worshipped a Sharke or some other Fish: the Hunter, a Lyon, Foxe, or other Beast, with many birds; the country-man, the water, and Earth' (Purchas, 1613: 728). Such references seldom failed to point out the errors behind such faith, but they pointed also to the natural power of strong creatures in the minds of such cultures. There is something ubiquitous at work here. It seems to be accepted, implicitly and from an early date, that the shark swam supreme in its natural element. Attracted by the alliteration, perhaps, John Barlow, in his exposition on St Paul's Second Letter to Timothy, described how 'like a pickrell in a pond, or sharke in the sea, the lesser fishes, deuoures the poorer sort, with a plausible, invisible consumption' (Barlow, 1625).

The flexibility and topicality of a new word also encouraged its use in translation. There is a straightforward use of the 'watry, Southwards sharke', possibly used for the rhyme, in Zachary Catlin's translation of Ovid's *Amores* (Ovid, 1639:C3). Arthur Gorges, turning Lucan's *Pharsalia* into modern English quatrains, again uses 'shark' alliteratively, and elusively, perhaps meaning 'struggle over':

> Neither could they so sharke and share
> The flesh, whereby the bones were bare
>
> (Lucan, transl. Gorges, 1614:303)

Other elegant employments of the word come in John Bidle's translation of Juvenal, and Thomas May's rendering of Martial's Epigram Against Cantharus, where sharking stands in for gluttony:

> When thou at others cost cat'st still,
> Thou roar'st, thou threaten'st, and speak'st ill.
> Such pride and fiercenesse sits not thee:
> Thou canst not sharke, and be so free
>
> (Martial, transl. May, 1629:F8)

It was of course an ideal English word to deploy in a 1623 translation of Mateo Alemán's *The rogue: or The life of Guzman de Alfarache,* when referring to a scavenging official who, out of self-interest, exceeds his duty and his powers (Alemán, transl. Murford, 1650:40). And it worked, too, in the 1626 translation of Boccalini's Italian political writings (Boccalini, 1626:101). The shark as politician also proves durable down the years.

Reaching maturity: 'shark' in the later seventeenth century

By the middle of the seventeenth century, the word 'shark' in its various meanings displays a settled maturity. This familiarity may, however, have evolved at the expense of earlier spontaneity as a vogue word: the freedom to apply it wherever the writer thinks fit may already be diminishing. More specific vogues, for phrases and compounds, are noticeable. The alliterative phrase 'to shift and shark', that is, to scavenge imaginatively, is found from the 1610s, but is fashionable around mid-century (Taylor [*Taylor's Revenge*], n.d.: A5). 'Shark' appears in one or two variant forms, as under the pen of the 'Fifth Monarchist' independent preacher John Tillinghast when reflecting on reluctant prophets: 'How did *Jonas* shark to put off that work God sent him about? and *Moses* little less at first, when God was sending him into *Aegypt*' (Tillinghast, 1658: 8). This is surely imaginative procrastination. But Tillinghast's sharking and other similar uses are excursions under a familiar theme. People 'shark away', in a sense of making themselves scarce (Duffett, 1678: 63). They also shark 'in corners', or 'abroad', conjuring in the reader's mind the act of sniffing out prey and prizes. Sharking, like other more elevated actions in life, is often competitive. If one shark fails in the bid to gull a victim, another will assuredly succeed. In the shark's pool, everything is a scramble for profit and preferment.

The human shark continued to be listed among the types best shunned by discerning gentlemen, a stock danger in advice literature (Brathwaite, 1652:110; J.W., 1698: 26 et seq). He (overwhelmingly, though not invariably, the predatory human shark is masculine) was also one of the snares confronting a would-be 'pilgrim' in the land of the living (Denny, 1653:196). Sharks were perhaps becoming more and more often, in contemporary usage, down-at-heel, base and poor, strangely vulnerable: in winter, 'at ten at night the shivering sharke makes musique with his teeth' (de Bergerac, 1658:5; Cavendish, 1671:396). Though he might pretend to a higher status, he was really nothing but a 'tater'd Bulley, or Lowsy Shark o'th Town' (Misonebulonides, 1682: single sheet). The latter usage is significant. Led by the source of print, discourse and news, the later seventeenth-century shark was an urban phenomenon. He was closely associated with fraud at the gaming tables, as can clearly be seen in a collection of 'the newest songs and merry catches' sung around London in the 1670s (W.P., 1677:75).

In the growing number of publications describing England's increasingly important Atlantic colonies, sharks are mentioned more frequently among the extraordinary things encountered in those waters (Blome, 1685:184). The shark affords one gloss for the 'wonders' referenced in Psalm 107 (Pell, 1659). These marvels are, like other branches of knowledge, in need of classification. During the years witnessing the birth of the Royal Society, the age of proto-science, when Hooke and Newton were active, one sees further attempts at a basic taxonomy, of 'scientific' efforts to distinguish species (Hubert, 1664:13): 'The Nurse is just like a Shark, only its skin is rougher, and is used for making the finest Rasps' (Dampier, 1697:25). Of course, attempts to order and classify move fastest in learned Latin works on natural history, particularly in the wake of the sixteenth-century Swiss naturalist Conrad Gessner's *Historiae Animalium*, but whether in Latin or in English it does not necessarily move very far or very fast until after 1700. For proof of this the reader might look into the catalogue of the Museum of the Royal Society at Gresham College in the 1680s, which lists specimens of 'white shark' and 'long-snouted shark', among other exhibits, but which also shows, by omission, the limited extent of such differentiation (Grew, 1685). Occasionally a species is given a fresh name, perhaps derived from foreign usage, as in this example of 1666:

> The *Requiem,* otherwise called the *Shark-Fish,* is a kind of *Sea-Dog* or *Sea-Wolf,* the most devouring of all Fishes, and the most greedy of mans flesh: He

> is much to be feared by such as go a swimming: He lives altogether by prey, and commonly follows the ships to feed on the filth cast out of them into the sea. These Monsters seem to be of a yellowish colour in the water: Some of them are of an unmeasurable length and bigness, and such as are able to cut a man in two at one bite: Their skin is rough, and there are made of it soft files to polish wood: Their heads are flat, and the opening of their mouth is not just before the snout, but under it: Whence it comes, that to fasten on their prey they are forc'd to turn their bellies almost upwards: Their teeth are very sharp and very broad, being jagged all about like a Saw: Some of them have three or four ranks of these in each jaw-bone: These teeth lye within the gums, but they make them sufficiently appear when there is occasion . . . Some Nations call this Monster *Tiburon* and *Tuberon:* But the *French* and *Portuguez* commonly call it *Requiem,* that is to say, *Rest*.
>
> (de Rochefort, 1666: 102)

Descriptions of a shark's behaviour remain fixed on its indiscriminate eating and scavenging. Sharks are cannibalistic: they routinely devour other sharks (*Poor Robins character*, 1672: 5). And they allegedly eat anything else as well. Citing the sixteenth-century Jesuit writer, naturalist and traveller Jose de Acosta, Alexander Ross reminds his readers that 'The Shark or tiburon is a strange fish, out of whose gullet he did see drawn, a butchers great knife, and great iron hook, and a piece of an oxes head, with one whole horn, their teeth are as sharp as rasors, for he saw Sharks leap out of the water, and with a strange nimblenesse snap off both the flesh and bone of a horses leg hanging upon a stake, as if it had been the stalk of a lettice' (Ross, 1652:216). Like the 'Pike, Barbel, Fork-fish, all stinging Water-Serpents, and hurtfull Fish', the shark is governed astrologically by the Planet Mars (Ramesey, 1653:55). It is 'an enemy to man' (Heyrick, 1691:19). Sharks continue to bite humans, notably two youths killed while swimming near their ship on Oliver Cromwell's expedition to the West Indies, the so-called Western Design of 1655 (I. S., 1655:11). Tiburons, however the word is used in English, share the customary characteristics and are at this point in time often if not invariably placed securely in Spanish usage (N. N., 1655).

Etymology confused the seventeenth-century writer as much as it confuses the modern student. Edward Terry, Chaplain to the English diplomat Thomas Roe on his voyage to India and the Mughal court, suggests that 'The *Shark* hath not this name for nothing; for he will make a morsell of any thing he can catch, master and devour' (Terry, 1655:8). A shark's a shark because it sharks!

The sailors' antipathy is by now thoroughly well established, and presented squarely as an expression of revenge for the shark's apocryphal or exaggerated attacks on fellow mariners, rather more frequently than as a creature of ill-omen:

> There is a Fish called a Sharke . . . a common enemy to Saylers and all others that venture, in Calmes, to commit their naked bodies to the sea (for he often bites off Legs, sometimes Armes, and now and then swallowes the

whole body, if the Fish bee great): So when the Saylers take them, they use them accordingly. Sometimes by putting out their eyes, and throwing them over bord; sometimes by mangling and cutting their bodies, finns, and tayles, making them a prey to others, who were mercilesse Tyrants themselves; And in this kind of justice they are very Accurate.

(Ligon, 1657:5)

In this parallel literal/figurative descent of the word, one senses that the latter is now the more frequent and the more important to common usage. Guy Miege's English-French dictionary of 1677 offers no equivalent to the word as applied to a fish, but provides a nice translation for the human shark: '*or sharking fellow,* un filoux, un vagabond, qui cherche à attrapper quècun' (Miege, 1677). An end of century canting dictionary gives us three useful definitions, side by side, again putting the fish in a subordinate place:

Shark,
c. a Sharper; also a Large voracious Fish.

Sharper,
c. a Cheat, one that Lives by his Witts.

Sharp,
subtil, ready, quick or nimble-witted, forward, of lively Apprehension; also Poor and Needy

(B.E., 1699:L1v)

As a creature of Mars, the ancient Roman god of war, 'shark' became an appropriate name for a naval vessel in the 1690s, albeit a small brigantine with a crew of thirty (*An exact list*, 1693). Nippy and predatory, the brigantine or the sloop is well fitted to the fish's commonly accepted characteristics. So far as one can tell, nobody seems to consider the name unlucky within a profession full of superstitions. In a curious 'parable' of the 1690s, the shark – perhaps for the first time – appeared in the title of a literary composition, as the principal protagonist in the cautionary tale of *The Shark and the Herring Pond* (*Monsieur in a Mouse-Trap*, 1691). This being political satire of a rather obvious sort, the 'shark' stands in for the French king Louis XIV: strong, crafty, unscrupulous, warlike, and a terror to his own citizens and other nations alike. Yet it is interesting to note that Louis is likened to the shark of the sea: imperial, if tyrannous and so distinctly un-British, in his realm. A contrast indeed to the 'land shark' who picks up the negative constructs only. There was a short-lived vogue for printing such 'parables', by way of commenting on current affairs. As the author says, he follows similar efforts centring on bear-baiting and magpies, and his shark parable was in turn critiqued by rival authors, for example by the equally anonymous source of *A lash for the parable-makers* (1691:4).

Established meanings

Much changes in the usage of this word through the seventeenth century, but many of the essentials do not change at all. John Kersey's *English Dictionary* of 1702, which set out to list words frequently in use, ventured only basic definitions: a blue shark is 'a fish', while the shark is both a 'greedy fish' and, rather circuitously, a 'sharking fellow'. Six decades on, things were not so different in Fenning's *Royal English Dictionary* of 1763. Here, the shark was defined in one composite entry as 'a ravenous sea-fish, which will sever a man in two at a bite. A sly greedy fellow. Trick; fraud.', while the verb means 'To pick up hastily or slily. Neuterly, to cheat; to trick' (Fenning, 1763). Shakespeare's use of 'shark up' in *Hamlet* was helpfully if rather spuriously defined in the 1740s as 'to pick up in a thievish manner', and was said to derive from the French 'chercher' (*Works of Shakespeare*, 1747: Glossary). Dr Johnson in mid century passed over the verb but gave three separate definitions of the noun. First, it meant 'A voracious sea-fish. "His jaws horrifick arm'd with threefold fate, The direful *shark*." *Thomson's Summer.*' Second, it represented 'A greedy artful fellow; one who fills his pockets by sly tricks. "David's messengers are sent back to him, like so many sharks and runnagates, only for endeavouring to compliment an ill-nature out of itself, and seeking that by petition which they might have commanded by their sword." *South's Sermons.*' Finally, a shark was a 'Trick; fraud; petty rapine. "Wretches who live upon the *shark*, and other men's sins, the common prisoners of youth, equally desperate in their fortunes and their manners, and getting their very bread by the damnation of souls." *South's Sermons.*' The great lexicographer here gives us little that is new. Thomson's *Seasons* and South's *Sermons* were familiar works, widely anthologised. But he points to the endurance of the land-shark alongside the ocean creature. While from this point we largely abandon him to his drier pursuits, we can never forget that the human shark remains a frequently encountered figure in English writing of all kinds, from that day to this.

For as long as the eighteenth century lasts, definitions accessible to a general readership seek no great subtlety. One word-list defined the shark as 'a great devourer' with 'three rows of teeth on each Jaw'. But, setting its scholarship in context, this work also defines the seal as 'a sort of Shark' (Wafer, 1704:203). Injections of classical references are perhaps more common than they were. The word is translated 'Canis, Carcharias' in a classical dictionary of around the same date (Ray, 1703:25). Some dictionaries give as an alternative, 'Lamia Rondeletii' (see Sibbald, 1710:50). And a greater distinction can be discerned between Latin and Greek equivalents: in a translation of Oppian's *Halieutica*, the White Shark is the Canis Carcharias, the 'soft shark' the Canis Mollis or Maltha, and the Blue Shark Glaucus (Oppian, 1722:226). As the natural world was grouped and standardised ever more precisely, there was a gradual trend towards greater, if still rudimentary shark classification. Off Norway, we are told in the 1760s, 'There is one species of the shark, called haec-moeren, ten fathoms in length, whose liver yields three casks of train-oil' (Smollett, 1768–9, i, 93). In the same decade John Berkenhout distinguished carefully between the white shark (carcharias) and the blue (glaucus) (Berkenhout, 1769–72: i, 63–4). It is the blue shark that frequents Cornish coasts during the pilchard-fishing season, destroying nets.

The London physician and scientist, and advocate of Linnaeus's taxonomical work, William Watson, read a precise observational paper on a blue shark, '*Squalus glaucus*', taken off the Devon coast, and provided a nice illustration in the Royal Society's *Proceedings* (*Philosophical Transactions* 68: 789) (Figure 2.3). Eventually, Linnaeus placed fifteen species within the wider shark-dogfish family (Pulteney, 1781:88). The eighteenth-century interest in finding and classifying fossils turned up many specimens of ancient shark across England. One tantalizing old question of classification necessarily remained unresolved. The debate over whether it was a shark or a whale that swallowed Jonah rumbled on endlessly, on both sides of the Atlantic, with neither side coming close to converting the other.

'Shark' remains imperfectly understood, defined by a limited range of instinctive habits and characteristics. Defiantly non-human, and 'foreign' itself, the wonder is that it translates so often and so flexibly onto perceived parallel human traits, almost all of them undesirable, but by this time its reputation has been firmly established in modern cultural representation. Can we avoid noticing that sharks fare worse through long association with the wide variety of rapacious or sly human behaviours, a strong indication about which of us is 'worse'?

References

Numerous primary works were consulted in the excellent resource that is Chadwyck-Healey Literature Collections online.

Adams, T. (1613) *Heauen and earth reconcil'd A sermon preached at Saint Paules church in Bedford, October. 3. 1612. At the visitation of the right Wor. M. Eland, Archdeacon of Bedford.* STC 122.

Anon. *A lash for the parable-makers* (1691) London. Wing L462A.

Bancroft, T. 1639. *Two bookes of epigrammes, and epitaphs Dedicated to two top-branches of gentry: Sir Charles Shirley, Baronet, and William Davenport, Esquire.* STC 1354.

Barlow, J. (1625) *An exposition of the second epistle of the apostle Paul to Timothy, the first chapter . . .* Cited from the epistle to the reader. STC 1434.

Belchier, D. (1618) *Hans Beer-Pot his inuisible comedie, of see me, and see me not. Acted in the Low Countries, by an honest company of health-drinkers.*

Berkenhout, J. (1769–72) *Outlines of the natural history of Great Britain and Ireland. Containing a systematic arrangement and concise description of all the animals, vegetables, and fossiles*, London.

Blome, R. (1685) *The English empire in America, or, A prospect of His Majesties dominions in the West-Indies . . . with an account of the discovery, scituation, product, and other excellencies of these countries.* Wing C7319.

Boccalini, T. (1626) *The new-found politicke Disclosing the secret natures and dispositions as well of priuate persons as of statesmen and courtiers; wherein the gouernments, greatnesse, and power of the most notable kingdomes and common-wealths of the world are discouered and censured.* London. STC 3185.

Brathwaite, R. (1615) *A strappado for the Diuell Epigrams and satyres alluding to the time, with diuers measures of no lesse delight. By Misosukos, to his friend Philokrates*. STC 3588.

Brathwaite, R. (1652) *Times treasury, or, Academy for gentry laying downe excellent grounds, both divine and humane, in relation to sexes of both kindes: for their accomplishment in arguments of discourse, habit, fashion and happy progresse in their spirituall conversation . . .*, London. Wing B4276.

Brathwaite, R. (1658) *The honest ghost, or, A voice from the vault*, London.

Philoeusebias Misonebulonides (1682) *A letter to Hilton, the grand informer in answer to his several late printed libels*. Wing 1705.

Cavendish, M. (1671) *Natures picture drawn by fancies pencil to the life being several feigned stories, comical, tragical, tragi-comical, poetical, romanicical, philosophical, historical, and moral*. Wing N856.

Collins, S. (1685) *A systeme of anatomy, treating of the body of man, beasts, birds, fish, insects, and plants illustrated with many schemes, consisting of variety of elegant figures, drawn from the life. . . .* Wing C5387.

Dampier, W. (1697) *A new voyage round the world. . . .* Wing D161/165.

de Bergerac, C. (1658) *Satyrical characters and handsome descriptions in letters written to severall persons of quality by Monsieur Cyrano de Bergerac*. Anon. translation. Wing C7718.

Dekker, T. (1609) *Lanthorne and candle-light. Or, The bell-mans second nights-walke In which he brings to light, a brood of more strange villanies than ener [sic] were till this yeare discouered*. London. STC 6486.

Denny, W. (1653) *Pelecanicidium, or, The Christian adviser against self-murder together with a guide and the pilgrims passe to the land of the living*, London. Wing D1051.

Dennys, J. (1613) *The secrets of angling Teaching, the choisest tooles baytes and seasons, for the taking of any fish, in pond or riuer: practised and familiarly opened in three bookes*. STC 6611.

de Rochefort, C. (1666) *The history of the Caribby-islands, viz, Barbados, St Christophers, St Vincents, Martinico, Dominico, Barbouthos, Monserrat, Mevis, Antego, &c*. Translated by John Davies. Wing R1740.

Dobson, E. (1644) *The declaration, vindication, and protestation, of Edward Dobson, citizen, and stationer, of London. Wherein is shewed the many illegall and unjust imprisonments which the said stationer hath suffered, through the malicious and envious informations of Brownists, Anabaptists, Antinomians, and other seditious sectaries . . .*, London. Wing D1751.

Duffett, T. (1678) *Psyche debauch'd a comedy*. Wing D2452.

E., B. (1699) *A new dictionary of the canting crew in its several tribes of gypsies, beggers [sic], thieves, cheats &c., with an addition of some proverbs, phrases, figurative speeches &c.: useful for all sorts of people (especially foreigners) to secure their money and preserve their lives*. Wing E4.

Earle, J. 1628. *Micro-cosmographie, or, A peece of the world discovered in essayes and characters*. STC 7440.2.

Edwards, S. K. (2009) Milton's reformed animals: an early modern bestiary: Shark. *Milton Quarterly*, 43.

Eilperin, J. (2011) *Demon Fish*, London.

An exact list of Their Majesties and the Dutch fleet, design'd for the year, 1693. (1693) Wing E3661.

Fenning, D. (1763) *The royal English dictionary: or, a treasury of the English language . . .*, 2nd edn, London.

Fitzherbert, J. (1523) *Here begynneth a ryght frutefull mater . . .* STC 11005.

Goodwin, J. (1642) *Impvtatio fidei. Or a treatise of justification wherein ye imputation of faith for righteousness (mentioned Rom: 43.5.) is explained & also yt great question largly handled. . . .* Wing G1172.

Gordon, R. (1625) *Encouragements. For such as shall have intention to bee vnder-takers in the new plantation of Cape Briton, now New Galloway in America, by mee Lochinvar*, Edinburgh. STC 12069.

Grew, N. (1685) *Musaeum regalis societatis, or, A catalogue and description of the natural and artificial rarities belonging to the Royal Society and preserved at Gresham Colledge,* London. Wing G1952A.

Hakluyt, R. (1625) *The Principal Navigations*, STC 12626a.

Hawkins, R. (1622) *The observations of Sir Richard Hawkins Knight, in his voiage into the South Sea. Anno Domini 1593*. STC 12962.

Herbert, T. (1634) *A relation of some yeares trauaile begunne anno 1626. Into Afrique and the greater Asia, especially the territories of the Persian monarchie: and some parts of the orientall Indies, and iles adiacent. Of their religion, language, habit, discent, ceremonies, and other matters concerning them.* STC 13190.

Herrick, R. (1648) *Hesperides*. Wing H1595.

Heyrick, T. (1691) *Miscellany poems*. Wing H1753.

Hubert, R (1664) *A catalogue of many natural rarities . . . collected by Robert Hubert, aliàs Forges, Gent. . . . and dayly to be seen at the place called the Musick House at the Miter, near the west end of St. Pauls Church*, London. Wing H3243.

Jones, T. 'The Xoc, the Sharke, and the Sea Dogs: An Historical Encounter'. http://www.mesoweb.com/pari/publications/RT07/Xoc.pdf (Accessed 13.05.2012).

Jonson, B. (1620) *Epicoene*. London.

Ligon, R. (1657) *A true & exact history of the island of Barbados illustrated with a mapp of the island, as also the principall trees and plants there, set forth in their due proportions and shapes, drawne out by their severall and respective scales*. Wing L2075.

Lodge, T. (1592) *Euphues shadow, the battaile of the senses*, London. STC 16656.

Lucan (1614) *Lucans Pharsalia containing the ciuill warres betweene Caesar and Pompey. Written in Latine heroicall verse by M. Annaeus Lucanus. Translated into English verse by Sir Arthur Gorges Knight. Whereunto is annexed the life of the authour, collected out of diuers authors.* STC 16884.

Martial (1629) *Selected epigrams of Martial. Englished by Thomas May Esquire*. STC 17494.

Mercurius cambro-Britannus, the Brittish mercury, or, The Welch diurnall communicating remarkable intelligences and true newes to awle the whole kingdome.

Middleton, T. (1608) *A Mad World, My Masters*, London. STC 17888.

Miege, G. (1677) *A new dictionary French and English with another English and French according to the present use and modern orthography of the French inrich'd with new words, choice phrases, and apposite proverbs*. Wing M2016.

Mill, H. (1639) *Poems occasioned by a melancholy vision or A melancholy vision vpon diuers theames enlarged which by seuerall arguments ensuinge is showed*. STC 17922.

Monsieur in a mouse-trap, or, The Parable of the shark and herring-pond by the author of The magpies (1691). Edinburgh. Wing M2459.

Murford, N. (1650) *Fragmenta poetica, or, Miscelanies of poetical musings, moral and divine by Nich. Murford*. Wing M3100.

N., N. (1655) *America: or An exact description of the West-Indies: more especially of those provinces which are under the dominion of the King of Spain*. Wing N26.

Nicholl, J. (1607) *An houre glasse of Indian newes. Or A true and tragicall discourse, shewing the most lamentable miseries, and distressed calamities indured by 67 Englishmen, which were sent for a supply to the planting in Guiana in the yeare. 1605,* London. STC 18532.

Oppian (1722) *Oppian's Halieuticks of the nature of fishes and fishing of the ancients in V. books. Translated from the Greek, with an account of Oppian's life and writings, and a catalogue of his fishes*, Oxford.

Ovid (1639) *Publ [sic] Ovid. De tristibus: or Mour nefull [sic] elegies in five bookes: composed in his banishment, part at sea, and part at Tomos, a city of Pontus. Translated into English verse by Zachary Catlin, Mr. of Arts. Suffolke*. STC 18981.

P., W. (1677) *The wits academy, or, The muses delight consisting of merry dialogues upon various occasions* . . . Wing P139.

Parker, R. (1607) *A scholasticall discourse against symbolizing with Antichrist in ceremonies: especially in the signe of the crosse*. STC 19294.

Pell, D. (1659) *Pelagos. Nec inter vivos, nec inter mortuos, neither amongst the living, nor amongst the dead. Or, An improvement of the sea, upon the nine nautical verses in the 107. Psalm*, London. Wing P1069.

Poor Robins character of a Dutch-man as also his predictions on the affairs of the United Provinces of Holland, together with a brief epitomy of the ingratitude of the Dutch, and the English at Amboyna, Polaroon and other islands in the East Indies (1672). London. Wing P2877.

Pulteney, R. (1781) *A general view of the writings of Linnæus*, London.

Purchas, S. (1613) *Purchas his pilgrimage*. STC 20505.

Purchas, S. (1625) *Pilgrimes*. London.

R., C. (1569) *The true discription of this maruellous straunge fish*, STC 20570.

Ralegh, W. (1614) *The History of the World*, London.

Ramesey, W. (1653) *Astrologia restaurata, or, Astrologie restored being an introduction to the general and chief part of the language of the stars: in four books*. . . . Wing R201.

Randolph, T. (1632) *The jealous lovers A comedie presented to their gracious Majesties at Cambridge, by the students of Trinity-Colledge*. STC 20692. Commendatory verse by Edward Fraunces.

Ray, J. (1703) *Nomenclator classicus, sive dictionariolum trilingue, Secundùm Locos communes. Nominibus usitatioribus Anglicis, Latinis, Graecis, ordine dispositis. A classical nomenclator with the gender and declension of each word and the quantities of the syllable*, London.

A relation of a terrible monster taken by a fisherman neere Wollage, July the 15. 1642. and is now to be seen in Kings street, Westminster . . . (1642) Wing R796.

Ross, A. (1652) *Arcana microcosmi, or, The hid secrets of man's body discovered in an anatomical duel between Aristotle and Galen concerning the parts thereof* . . . Wing R1947.

Rowlands, S. (1609) *The knave of clubs*. STC 21387.

Rowley, S. (1605) *When you see me, you know me. Or the famous chronicle historie of King Henry the eight, with the birth and vertuous life of Edward Prince of Wales*. STC 21417.

S., I. (1655) *A brief and perfect journal of the late preceedings and successe of the English army in the West-Indies, continued until June the 24th 1655* . . ., London. Wing S35.

S., S. (1616) *The honest lawyer Acted by the Queenes Maiesties Seruants. Written by S.S.*

Saunderson, R. (1627) Ten sermons preached I. Ad clerum. 3. II. Ad magistratum. 3. III. Ad populum. 4. By Robert Saunderson Bachellor in Diuinitie, sometimes fellow of Lincolne Colledge in Oxford. STC 21705.

Sibbald, R. (1710) *The history, ancient and modern, of the sheriffdoms of Fife and Kinross; with the description of both, and of the firths of Forth and Tay, And The Islands in Them . . .*, Edinburgh.

Smollett, T. (1768–9) *The present state of all nations . . .*, London.

Sydenham, H. (1626) *Iacob and Esau. Election. Reprobation Opened and discussed by way of sermon at Pauls Crosse, March 4. 1622*. STC 23567.

Taylor, J. (1622) *The water-cormorant his complaint against a brood of land-cormorants. Diuided into fourteene satyres*. STC 23813.

Taylor, J. (1612) *The sculler rowing from Tiber to Thames with his boate laden with a hotch-potch, or gallimawfry of sonnets, satyres, and epigrams. With an addition of pastorall equiuocques or the complaint of a shepheard*. STC 23791.

Taylors goose (1621) STC 23799.

Taylor, J. (n.d.) *Taylors revenge, or, The rymer William Fennor first . . .*, STC 23804.

Terry, E. (1655) *A voyage to East-India. Wherein some things are taken notice of in our passage thither, but many more in our abode there, within that rich and most spacious empire of the Great Mogol. Mix't with some parallel observations and inferences upon the storie, to profit as well as delight the reader*, London. Wing T782.

Tillinghast, J. (1658) *Elijah's mantle: or, The remaines of that late worthy and faithful servant of Jesus Christ*. Wing T1172.

Virgil's Bucolicks Engished [sic]. VVhereunto is added the translation of the two first satyrs of Iuvenal. By Iohn Bidle (1634). London. STC 24821.

W., J. (1698) *Youths safety: or, Advice to the younger sort, of either sex More valuable than gold. Laying open the wicked practices of the town-shifts, sharpers, sharks, beau's, sweeteners, rakes, intreaging town-jilts, to cheat, ruin and disgrace gentlemen, shopkeepers, apprentices, gentlewomen, servant-maids, &c*, London. Wing W76.

Wafer, L. (1704) *A new voyage and description of the isthmus of America. Giving an account of the author's abode there, The Form and Make of the Country, the Coasts, Hills, Rivers, &c. Woods, Soil, Weather, &c. Trees, Fruit, Beasts, Birds, Fish, &c . . .*, London.

Wood, W. (1634) *New Englands prospect – A true, lively, and experimentall description of that part of America, commonly called New England: discovering the state of that countrie, both as it stands to our new-come English planters; and to the old native inhabitants. Laying downe that which may both enrich the knowledge of the mind-travelling reader, or benefit the future voyager*, London. STC 25957.

Working, L. (2015) 'Savagery and the State: incivility and America in Jacobean political discourse', Durham PhD dissertation.

3 Sharks in poetry

Poets have been powerful creators of languages throughout human civilization: invested with the responsibility for preserving and transmitting values, beliefs, social codes, histories and the widest possible range of human experiences. Through a command of rhythms, rhetorical patterns and sound schemes, poets persuade. They harness the most powerful tools in any language and in any culture for forging and changing attitudes. It is, therefore, especially interesting to explore how poets have shaped sharks and their reception through the centuries.

We know from the 1569 broadside that woodcuts of sharks were in the public domain. There are certainly numerous representations of sea monsters in maps of the period, some of which are equipped with large teeth, although they otherwise do not conform to recognizable representations of sharks (Duzer, 2013). However, the Early Modern poets seem to have had little use for sharks. Instead there are many representations of land-sharks, usurers, lawyers, and scroungers. For instance, in 'A Light Heart's A Jewell' (Anonymous, 1602–1646 EBBA30152) the shark is a someone who doesn't repay his debts. In 'Here's To Thee Kind Harry' (to the tune of Here's to thee good Fellow) (Anonymous, 1627, EBBA 20203), the shark is a scrounger who never pays for his own drinks, even when he has money. In 'Folly Plainly Made Manifest' (Anonymous, 1671–1702, EBBA 21784) the husband complains that he spent all his money in the tavern and that now he is regarded as a shark, here meaning scrounger, and has been deserted by his false friends. It is really a version of the popularly adapted biblical story of the prodigal son (Luke 15:11–32). These ballads are typical of the representation of sharks, that is, land-sharks.

Subsequently, the real shark was to play a very significant role in representing human experience. In the early nineteenth century there was a significant increase in references to sharks in poetry. With the expansion of the empire and increased sea traffic and trade came an increased acquaintance with, and interest in, sharks. It was also an age of war. Naval wars were illustrated and represented in gruesome detail, but not usually their aftermaths. It was chiefly sharks that cleared up the casualties strewn across the watery battlegrounds. This cannot have been a pleasant sight for those present, or welcome news to loved ones back at home. The idea of the sailor becoming a shark meal, having served his country, would have been horrible. The poetry of the period reflects this tendency to spare the public full detail. However, there is also a broader representation of sharks than hitherto seen in poetry. There are several humorous representations, a focus on their relationship to the slave trade, their curious uses by other cultures, biblical analogies taken from Exodus to Job, instances of sharks being attacked, and even frightened sharks. In one of the more bizarre references, in Philip Gilbert Hamerton's *The Isles of Loch Awe* (1855), the shark bite is used to convey the pain of getting cramp while swimming.

Between 1950 and 2000, sharks proliferate in poetry. Aside from the by now more familiar references to shark jaws, we find much more imaginative representations of sharks, from soap in a bathtub, Robert Graves's *I wonder what it feels like to be drowned?* (1917), to a bath toy, Mark Cox's *Get me Again* (1998), a pool shark cleaner, Michael Hofmann's *Freebird* (1993), and even some sympathetic representations of the plight of sharks. In Hilda Doolittle's *The Walls Do Not Fall* (1944), it is the shark jaws of life that define who we are, while Eve Merriam's *Pentagonia* (1970) calls on us to love the shark for its sheer sharkiness. There are also the first mentions of shark repellents, from jackets in Anne Waldman's *Letter to Miss Idona Hand* (1989) to flowers in Edward Field's *Shark Island* (1992), and a keener focus on shark behaviour. Ted Hughes has a quick fix for a headache in his cautionary poem *The Mermaid's Purse* (1995). The 'mermaid's purse', one term used for a shark egg sack, becomes a bifurcated pun of a mermaid and her purse. The mermaid, seeking a remedy for her headache, reaches into her purse for an aspirin but instead gets her head bitten off by a shark. Thom Gunn's *Cannibal* (1992) offers a profile of a shark that begins with its mouth like knives, unusually, tucked out of sight. Again the shark is represented as ruthless through an analogy with human relationships. In this short poem, one of *Three for Children,* the shark is described not merely as eating his brother but very violently biting him 'into bits', referring to the fact that some shark species do indeed eat their siblings in the womb (referred to as ovophagy). Gunn makes the shark's behaviour appear gruesome but it is simply an efficient means of minimizing waste and keeping shark populations down naturally.

There is a very wide range of representation of sharks in poetry throughout the ages, some of which can be discussed in terms of broad categories, such as their appropriations for the purposes of political protest through to the horrors of the slave trade and the fear of shark attack.

Political protest and sharks

An early reference to shark's teeth appears in Andrew Marvell's *Last Instructions to A Painter* (1667). Here, Marvell complains about the taxes that are enforced on English goods, rendering them uncompetitive with the Dutch at a time when England is trying to battle for supremacy in sea-trading. For Marvell, this tax or 'excise' on commodities is a 'monster' – one that he considers is even more deadly than the toothy shark.

With hundred rows of Teeth the Shark exceeds,
And on all Trades like Casawar she feeds;
Chops of the piece wheres' e're she close the Jaw,
Else swallows all down her indented Maw.

(135–138)

The fused image of a cassowary and a shark enables Marvell to incorporate land and sea in this tax system that cripples or swallows everything it comes into contact with. The idea of a monster with hundreds of rows of teeth more than the shark is clearly intended to convey the inescapability of this lethal tax system (Figure 3.1). As Noelle Gallagher pointedly remarks, Marvell's take on this confrontation with the Dutch was that it was 'a haphazard series of blunders perpetrated by corrupt, incompetent officials' (Gallagher, 2011:89). Through several of the poems we will discuss, sharks are referred to as voracious, or greedy, even though all biological evidence points to the contrary (Domeier, 2012). The wide mouth of some of the larger predatory sharks is probably the reason for falsely supposing them greedy creatures. The slicing motion of the jaws of both the great white and the tiger sharks may indeed give the impression that they are filling their huge mouths, when they are merely taking a bite – unlike the men at the centre of Marvell's protest.

Thomas Heyrick's *The Submarine Voyage* (1691) is a response to the political and religious squabbling of his day. In this expansive exploration of the oceans he seeks to put problems at home in some perspective and elevate minds to nobler ambition (Houston and Pincus, 2001). However bold his aspirations for poetry, his representation of sharks is not generous. Sharks are described as lurking around ships waiting for them to be wrecked:

The Shark, an Enemy to Man,
That craftily about the Ships doth stay,
And never Spares his Prey:
Seales, that in hollow Caves delight,
And shun Man's Dangerous Sight,
On Barren Rocks and Isles are bred,
Where foot of Man did never tread.
The Remora, the Wonder of the Sea,
That Ships even under sail can stay:
Small in his Bulk, but hoisting round their Keels,
No Waves or Tydes the Captive force away:
Whom Neptune did forbid to touch his Chariot-wheels.

(566–577)

This representation of the shark is surprising given Heyrick's acknowledgement at the beginning of the book that these wild animals kill for food, unlike the fools at home who kill each other over trifles. Even though the poem contextualises the shark with references to crocodiles, tempests and rocks, it cites only the shark as

'Enemy to Man'. While the shark is vilified for being smart and sticking close to the boat, the humble remora is celebrated as the 'Wonder of the Sea' in being able to stick to anything, even to a ship that is travelling at speed. This sticking power of the remora is harnessed as a positive analogy for human stamina and fortitude, perhaps even loyalty to superior power.

The death of the exiled King James II of England in 1701 drew a range of interesting elegies from notable writers, but in poems such *The British Muse* (1702) by John Tutchin he is described as no less than a cannibalistic shark.

> As in the Waters we do Fishes find,
> Which do devour and prey upon their Kind;
> This Princely Shark on his own Species fed,
> When Cause requir'd, and Rome the Order made.
> Thus Coleman to his Jaws a Victim fell,
> Sent in a Jugler's Box to plot in Hell.
> Vain Wretch! who could so fatally believe
> A Man enclin'd by Nature to deceive.
> With him what Wretches would the Scepter trust,
> And blasphemously call him James the Just?
>
> (77–86)

The 'Jaws' that make victims of men such as Coleman, here specifically mentioned, are fused with the mouth of Hell, so that any elegy to James becomes a blasphemy. England was still a Protestant nation and when James succeeded his brother Charles II as King in 1685 he was regarded with some suspicion on account of his Catholic faith. Here in this poem he is the puppet of Rome (the seat of Catholic power). James was soon ousted and William III and Mary II succeeded him as Protestant monarchs in what came to be known as The Glorious Revolution (1688), in which Tutchin himself participated.

Rather than call James a traitor, Tutchin chooses to call him a Princely Shark, as a devourer of his own people, a particularly gruesome and bloody metaphor to bring his readers closer to the true nature of his treachery, though clearly contributing to the misrepresentation of shark behaviour. Since accounts of savage cannibals were readily available in the public domain, the use of cannibalistic sharks is particularly interesting, and very probably a reflection of his personal observations of political relationships more generally. According to Peterson, Tutchin exposed many high-level misdemeanours through his pamphlets and poems (Peterson, 1964). Such noble intentions were not likely to win him many powerful friends and he ended his days in prison.

Herman Melville, better known for his novel *Moby-Dick* (1851), wrote a poem entitled *The Maldive Shark* (1881) based on his experience of being at sea. In this short poem the shark is described in tandem with the pilot fish. These pilot fish are the tiny creatures that usually swim immediately in front of sharks, giving the impression that they are guiding them. A more likely theory is that they seek scraps by staying close to sharks' teeth (Webster, 2003). In Melville's poem the shark is understood as being both unemotional and dumb – a set of swimming jaws:

> From his saw-pit of mouth, from his charnel of maw
> They have nothing of harm to dread,
> But liquidly glide on his ghastly flank
> Or before his Gorgonian head;
> Or lurk in the port of serrated teeth
> In white triple tiers of glittering gates,
> And there find a haven when peril's abroad,
> An asylum in jaws of the Fates!
> They are friends; and friendly they guide him to prey,
> Yet never partake of the treat—
> Eyes and brains to the dotard lethargic and dull,
> Pale ravener of horrible meat.
>
> (5–16)

Melville is more impressed by the pilot fish, and their relationship with a creature that all others fear. They appear to be sentinels to the 'white triple tiers of glittering gates', which in turn offer them protection. Not known for his subtlety, Melville here appropriates the shark to critique the political leadership of his day. Great power can be led by a small few, who are quick to retreat from danger. Kathleen Rooney suggests that the poem is a warning against capitalism: 'The pilot fish can be read as managerial workers in the way they actively lead the shark to its prey, selling out other defenceless fish, acting in sympathy with the predatory institution instead of with their fellow smaller fry' (Rooney, 2014:70151). While Melville paints a vivid picture of the pilot fish his representation of the shark is entirely negative, misrepresenting the creature to reflect human stupidity and greed.

Love and sharks

Love has been a focus for poets through the ages, and there is certainly a long tradition of resorting to the natural environment to speak of love's passions. In the *Nereides* or *Sea Eclogues* (1712) of William Diaper, the long pastoral poetical tradition, reaching back to Theocritus (*c*. 270 BC), is harnessed and recontextualised in an oceanic rather than a garden setting, although he was by no means the first to do this. Jacopo Sannazaro had turned pastoral shepherds into fishermen in his *Eclogae piscatoriae* (literary fish) in 1526 (Passmann and Hermann, 2005:43). In Diaper's *Eclogue 1*, Glaucus attempts to convince Cymothoe that he is in love only with her, offering the hunting of a shark in a string of memories of bold adventures that are intended to convey his fearlessness in a way that is also suggestive of the sexual conquest of Cymothoe herself. Although this is the first instance of the idea of shark hunting as a sign of fearlessness in English poetry, it will become a much more common theme in a wider variety of subsequent literary representations. Diaper's transformation of a greenscape to bluescape was not well received in spite of nereides, or mermaids, having enjoyed considerable popularity in English culture for centuries (Weinreb and Hibbert, 1983). His *Eclogue XIII* (1712) describes the shark as the toothy predator that has become an iconic representation of the dangerous ocean.

The Shark with pointed Teeth is arm'd for Prey;
He breaks thro' all, and clears the liquid Way;
While the fond Sucking-fish (a harmless Breed)
With fastned Lips supply their daily need,
And with a Mouth unarm'd they clinging feed.
No Lovesick Nymph's, or wanton Triton's Kiss
Is half so lasting, or so close as his.

(47–53)

Diaper's poem references Milton's *Paradise Lost.* In Book VIII Milton's Raphael (the sociable angel) tells Adam (the father of mankind) that his happiness lies in obeying God. He warns Adam to seek only knowledge useful for husbanding Paradise and not to waste his time on idle speculation or things that don't concern him. This is summed up in Raphael's advice: 'Seek to know no more', which Diaper's poem here repeats (105). In the *Nereides* wisdom is derived from observing nature. The shark is described merely as being armed with pointed teeth and a body that slices through the ocean. Interestingly, more space is given to the remora, which picks off parasites and picks up scraps from the shark's meals. The clear reason for this fulsome attention to the remora is that its front dorsal fins have been adapted into a complex disc that can, through suction, attach to sharks, thereby suggesting to the poet a pair of kissing lips (Beckert, Flammang et al., 2015). By contrast with the shark, the remora has a 'Mouth unarm'd'. The analogy with the lovesick Nymphs or Triton's Kiss creates the final image of the shark and its remora as a loving couple. Diaper's aim is not to make the shark terrifying but, rather, to indicate a more positive side to its powerful nature as it travels in a symbiotic relationship with its remora in a perpetual state of being kissed.

In *Woman! when I behold thee flippant* (1817) John Keats uses a simile of a greedy shark, invoked in order to convey the speaker's hunger for his lover's words. He remarks upon her beauty in Petrarchan fashion and then concludes that it is easily forgotten when she speaks:

These lures I straight forget,—e'en ere I dine,
Or thrice my palate moisten: but when I mark
Such charms with mild intelligences shine,
My ear is open like a greedy shark,
To catch the tunings of a voice divine.

(24–28)

Here, the shark's open mouth becomes the ear of the poet, greedily devouring words. Greediness is the key attribute of sharks in the early nineteenth century. Keats' choice of the shark rather than 'fluffier' creatures that lend themselves more readily to a love poem gives the poem a potent animality and urgency. Sharks remain a rarity in love poetry of any age, although Richard Cobbold uses sharks to berate lovers who make their mistresses jealous for sport in *Fatal Jealousy* (1827). Here, the shark keeps company with the eagle for its ruthlessness and rapacity:

> Does eagle tear from bleeding lamb the heart?
> Does life-blood from the wounded bird depart?
> Does shark rapacious seize the swimming prey?
> More cruel man, who trifles in this way.
>
> (23–26)

The tearing and seizing of the eagle and shark are chosen to represent the pain of human jealousy. However, here in the poem the eagle and the shark behave this way because they must, whereas the lover is truly cruel simply because he enjoys it. In a later poem, Carol Ann Duffy will appropriate the shark to explore the emotional processes that manifest as being *Jealous as Hell* (1998). In Cobbold's poem the shark is used to epitomise cruelty, yet a cruelty that can still be surpassed by a lover.

Slave ships and sharks

Throughout the history of poetry in English, sympathetic representations of sharks have been rare. From the 1680s sharks begin to emerge from the metaphorically muddied waters of fusion with usurers, lawyers or scroungers. We also see a greater emphasis on shark attacks. Mariners' tales of enormous fish no doubt excited the imagination of those who heard them and were variously repeated and adapted to mythic proportions of monstrosity in the popular culture (Rappoport, 1928; 2007). From a wide survey of the poetry we can discern that sharks have a special relationship with sailors and boats. While some poems tell of unfortunate incidents when sailors forgot they were in wild environments and leapt from their boats without looking, other poems explore the contexts in which men are systematically expelled from boats. The most poignant of these poems relate to the slave trade. The poets' vociferous protestations contributed to the changing of attitudes and the eventual banning of slave trading in the United Kingdom in 1833.

In his very popular poem *Summer* (1730), James Thomson imagines the great navigators who must have been subject to challenging seas. In the midst of a tempest, its most dreadful creature emerges, lured by the stink of disease and death that often denoted the slave ships:

> Increasing still the terrors of these storms,
> His jaws horrific arm'd with threefold fate,

Here dwells the direful shark. Lured by the scent
Of steaming crowds, of rank disease, and death,
Behold! he rushing cuts the briny flood,
Swift as the gale can bear the ship along;
And, from the partners of that cruel trade,
Which spoils unhappy Guinea of her sons,
Demands his share of prey; demands themselves.
The stormy fates descend: one death involves
Tyrants and slaves; when straight, their mangled limbs
Crashing at once, he dyes the purple seas
With gore, and riots in the vengeful meal.

(1013–1025)

What elsewhere has been described as a 'fell triple saw' is here a 'threefold fate'. It is not clear why three has become the standard for descriptions of sharks' jaws, which can contain up to seven rows of emerging replacement teeth, except perhaps as an echo of the highly influential description of Milton's hellish threesome in *Paradise Lost*. Here, the shark first 'cuts' the sea to get to the ship, foreshadowing the fate of those who are tossed overboard in the tempest, and then gets a cut in the slave trade as disease takes hold on the crowded ship and the bodies are disposed of. The tempest takes on the role of a divine intervention that metes out death, the great leveller, to all the men on the ship, be they masters or slaves. Although the sharks eat indiscriminately, death at sea is a divine mercy to the slaves, circumventing worse fates ashore.

A century and a half later, John Greenleaf Whittier begins his *The Slave Ships* (1894) epigrammatically with two lines from Milton's *Lycidas,* an elegy written for one of his college acquaintances, Edward King, who was drowned at sea and whose body was never recovered. In Whittier's poem the slave ship is the 'fatal and perfidious bark' or treacherous ship. The poem relates how those who died in the damp and overpopulated cells were brought up first to be tossed to the sharks. Here the captain instructs his sailors as though they command the sharks. Sharks obediently appear beneath the ship, waiting patiently with their mouths open like cartilaginous trash cans for the ship's waste:

Beneath, the lean shark lay,
Waiting with wide and bloody jaw
His quick and human prey.

(46–48)

While there have already been plenty of references to sharks' insatiable appetite and keenness to eat human beings, here the shark is 'lean', suggesting real hunger. In this poem a symbiotic relationship is set up between the slave ships that must be cleansed and the sharks that must be fed. However, the fact that these slaves are still alive when the sharks start eating them adds another gruesome dimension to the slave traders' operations.

Voices of agony and blood,
From ship-deck and from sea.
The last dull plunge was heard,
The last wave caught its stain,
And the unsated shark looked up
For human hearts in vain.

(51–56)

The poem is based on the misadventure of the ship *Le Rodeur*, which in April 1819 was transporting slaves from Bonny in Africa when many of the men fell sick. They were disposed of, the poem suggests, without mercy. Interestingly, the sharks are not sensationalised as monsters but as a useful aid to the captain as he deals with the problem. The lack of emotion displayed by the captain who orders living men overboard puts him on a par with the sharks in the poem.

Robert Southey's *Joan of Arc* (1838) provides a more unexpected analogy, between Talbot and the shark. When Shakespeare writes the story of Joan of Arc in *Henry VI Part I*, drawn from Holinshed's Chronicles, he makes Talbot the hero of a play without heroes. Here, Talbot sees his son in one-to-one combat with Joan, and finally felled by her. Fastolffe flees, only to be confronted and finally cut down by Joan, before Talbot senior re-enters the scene to take control.

Fastolffe, all fierce and haughty as he was,
False to his former fame; for he beheld
The Maiden rushing onward, and such fear
Ran through his frame, as thrills the African,
When, grateful solace in the sultry hour,
He rises on the buoyant billow's breast,
And then beholds the inevitable shark
Close on him, open-mouth'd.

(501–508)

What is interesting about this passage is the imitation of the Miltonic epic simile that suddenly cuts away to a scene in which an African slave is floating in the open ocean to be seized by a shark. This kind of simile serves to heighten the drama of the present moment by taking the reader to another moment of high drama. The 'inevitable shark' means inevitable death when encountering a shark, though we know that this is not usually the case. In fact, Fastolffe is killed by Joan because he fears her.

Celia Thaxter was born and grew up by the sea, and much of her poetry reveals her imaginative connectedness to seascapes. In her poem *The Cruise of the Mystery* (1896) children are walking along the beach with their grandmother when they spot a ship sailing rapidly. The grandmother instantly recognises it as the ghost ship *Mystery*. She then proceeds to tell them the story of this slave ship being hit by a hurricane. She describes the slaves as 'herded cattle' (46) packed tightly together. The captain orders the men to throw the 200 dead slaves overboard, where sharks are waiting:

> From weltering wave to wave they tossed.
> Two hundred corpses, stiff and stark,
> At last were in the distance lost,
> A banquet for the wandering shark.
>
> (57–60)

As so often, the misfortune of those on board a sinking ship is represented as a good day for sharks. However, it does make clear that these are not shark attacks. The sharks get the corpses.

James Montgomery's *The Ocean, Written at Scarborough, in the summer of 1805* is one of the most vehement anti-slavery poems of the period. The Jamaican rebellion of 1831 led to the Slavery Abolition Act in 1833 which Montgomery celebrated with a series of songs (Montgomery, 1880). In this poem the shark awaits those who through avarice travel across the seas to pursue other nations to feed the slave trade in England. The poet believes God deliberately separated nations by the oceans and that punishment, in the form of sharks, awaits those who cross the seas:

> The shark hears their shrieks, and, ascending to day,
> Demands of the spoiler his share of the prey.
> Then joy to the tempest that whelms them beneath,
> And makes their destruction its sport!
> But woe to the winds that propitiously breathe,
> And waft them in safety to port,
> Where the vultures and vampires of Mammon resort;
> Where Europe exultingly drains
> The life-blood from Africa's veins;
> Where man rules o'er man with a merciless rod,
> And spurns at his footstool the image of God!
>
> (62–72)

Montgomery appears to be suggesting that being shipwrecked and eaten by sharks is a better fate than the slave trade that awaits these passengers when they reach port. The slave trade can exist only in a world turned upside down, where the image of God is brought low. The poem ends with a prophetic plea to remember what makes human society humane:

> —Let Ambition, the sin of the brave,
> And Avarice, the soul of a slave,
> No longer seduce your affections to roam
> From Liberty, Justice, Religion, AT HOME.
>
> (159–162)

In *Similes for Two Political Characters of 1819*, Percy Bysshe Shelley describes two political figures, namely Viscount Sidmouth, who was the home secretary, and Viscount Castlereagh, who was the foreign secretary, who both supported the Peterloo massacre of civilian protesters in Manchester on 16 August 1819. Shelley was so angered by it that he wrote this poem as a response (Mullan, 2013). To lend his full force of disapproval, Shelley again draws on the image of sharks lurking beneath slave ships, waiting for the sick and the dead to be thrown overboard:

> As a shark and dog-fish wait
> Under an Atlantic isle,
> For the negro-ship, whose freight
> Is the theme of their debate,
> Wrinkling their red gills the while—
>
> (11–15)

Perhaps even worse than the waiting sharks is the dehumanisation of the slaves as mere 'freight'.

Fear and shark attack

In the process of describing the dreadful conditions of slaves at sea, poets have certainly played their part in contributing to our fear of being attacked and eaten by sharks. They rarely consider the challenging lives of sharks in the oceans. As masters of language, poets are able to take an experience and turn it into our worst nightmare through skilful representation, and we need to be alert to the way our thinking about sharks is being guided.

John Scott's *An Essay on a Painting* (1782) is a study in poetical and visual representation. The key work in his discussion is John Singleton Copley's *Watson and the Shark,* which is given a full treatment in our discussion of the visual representation of sharks later in this book. In this section of the poem, Scott explains the importance of telling a tale coherently so that the viewer can see the processes of time and thought and motion:

> When Copley's Youth, swift-struggling thro' the wave,
> The anxious boatmen strain each nerve to save;
> As strives the ravenous shark to reach his prey,
> One lifts the javelin to arrest his way;
> And now, as near his dreadful jaws expand,
> One casts the cord, and one extends the hand:
> What care, what pity, mark their eager eyes!
> What hopes, what terrors, in our bosoms rise
>
> The skilful Painter, at whose option lie
> Positions various, fails not all to try;

And those prefers, where every part the best
Accordance keeps, illustrating the rest.

(317–328)

For Scott, Copley's painting is a drama depicting the struggle of Watson to get to the boat, the struggle of the shark to reach him and the struggle of the sailors to prevent the two from meeting. From these struggles the emotion of the moment is invoked in the viewer. The poet's image of the 'ravenous' shark opening its jaws just as the helping hand reaches to Watson suggests that the cooperative efforts of the sailors win the day, but the painting actually reveals otherwise. Watson is still in the water in the painting, and Copley suggests that he has lost a limb by cutting through his leg with a sharp water line.

In his *Bryan and Pereene* (1767), James Grainger tells the story of two lovers, a boy of twenty years old who has been in England for over a year, and a West-Indian lover. Just as his ship is coming to shore he sees Pereene waving a handkerchief at him. So eager is he to reunite with her that he will not wait for the ship to anchor and jumps into the sea. He swims towards the shore only to be bitten in two by a shark in full sight of his lover, who then dies of shock.

Then through the white surf did she haste,
To clasp her lovely swain;
When, ah! a shark bit through his waste:
His heart's blood dy'd the main!

He shriek'd! his half sprang from the wave,
Streaming with purple gore,
And soon it found a living grave,
And ah! was seen no more.

(41–48)

The ballad is reportedly based on a true story of a shark encounter on St Christopher Island. What is interesting here is that although the description of the boy's body streaming with blood is quite graphic, it is not sensationalised. The shark finishes its meal and quickly disappears, never to be seen again, while the boy is imagined in his living grave. Robert Charles Dallas's *Amabel* (1823) is also based on a true story of a shark attack in St Vincent in the Caribbean. A group of sailors are waiting for the wind to pick up so they can set sail, so they decide to have a swimming race. Frank is the first to reach the boat, then suddenly the mirth turns to anxiety as they shout to him three times, 'Haste! gain the boat'. The sailors have seen a shark approach. The terror felt at the sight of the shark is indicated through the desperate clutching at descriptions to convey how the 'monster', 'Leviathan', 'glutton shark', 'finny wolf with horrid jaw/ . . . arm'd with Death's fell triple saw' approaches. The three layers of teeth as three deadly saws are mirrored in the description of the shark's movement as 'lightning thro' the deep sea breaking' (157) to convey the zigzag ripple that is created by its dorsal fin, conveying the sense of its dangerous jaggedness. It is first described as a distant speck but is suddenly in their midst.

> Emerging half his mountain-hide,
> The monster strikes the trembling tide;
> And as he sidelong quivers through,
> His prey he singles from the crew—
> From off the Cutter's deck is seen
> The human form his jaws between;
>
> (173–178)

The poem describes the shark as moving sidelong towards a chosen victim. This is an accurate depiction of one of the shark's many strategies for getting a closer look at something in the water. Sideways-on is not typical shark hunting strategy, but rather registers curiosity and wariness. The poem creates the sense of an instantaneous attack by leaving the actual attack out. One minute the shark is sideways-on at a distance, and in the next Frank is in its jaws. Here, the shark carries Frank off into the depths. The sailors who watch in vain as Frank's body disappears from view are depicted as being caught in wonder. No one at this stage knows who the victim is and as the other swimmers head towards the boat they cast about to check who is present, and who is missing.

> As to the boat the swimmers ply,
> In silent search turns every eye,
> And soon, too soon, each voice is free—
> The general groan proclaims, 'Tis he!
>
> (187–190)

The poem ends with Amabel being told of Frank's demise and losing her mind. She eventually dies in sorrow with Frank's name on her lips. What is intriguing about this poem is the range of emotion, from the impatience of the sailors, who swim playfully, apparently indifferent to the shock of witnessing the attack, and Amabel's misery upon losing Frank. The poem chooses fatalism over sensationalising of the shark as it carefully selects its target: 'His prey he singles from the crew' (176).

Mary Botham Howitt's *Southern Seas* (1834) opens with a call to explore the seas with courage and to not be deterred by storms, rocks or sharks, and in Robert Southey's *The Curse of Kehama* (1838), the frightened shark flees the hero:

> The hungry Shark, at scent of prey,
> Toward Ladurlad darted;
> Beholding then that human form erect,
> How like a God the depths he trod,
> Appall'd the monster started,
> And in his fear departed.
>
> (10–15)

The shark is described as hungry but instantly recognizes the fearlessness of Ladurlad and flees. This section of the poem is indebted to Milton's *Paradise Lost* in which the ultimate adversary of God and Man perceives Adam's naked majesty and is awestricken (PL IV. 288–289). Even though this is a frightened shark it is still represented as a monster, simply for the purposes of making Ladurlad appear heroic.

In the following poem by George Croly, this fear of sharks infects a sailor's dreams. *The Modern Orlando* (1846) Canto V recounts what it is like to be asleep on a ship and awoken by a shark nightmare.

> Night drops death-dark; and, if on board a ship,
> At anchor in some Windward island's lee,
> You're sure to dream of taking your last trip,
> Down to the bottom of the 'deep deep' sea;
> A million sharks upon you making free!
> You feel their triple grinders taking bites!
> You scream, and bounce from bed; the bell strikes three;
> (The blackest hour of black West Indian nights;)
> You find the crew all up, and hammering the dead-lights!
>
> (73–81)

The biggest concern of any sailor is not to wreck the ship, but in this nightmare the consequences of a wreck are made sufficiently vivid to wake the sailor with a start. Once a sailor's body has hit the bottom of the sea, it is unlikely that he will still be alive, but in the dream the sailor is aware of sharks' teeth biting into him. The screaming and bouncing suggest the first moments of arousal from a nightmare and mimic the disorientation of being lost in the sea. It is not until the bell strikes that the sailor is fully aware of where he is, surrounded by his vigilant mates. While the 'dead lights', referred to in this poem, can simply mean deck prisms, which were designed to reflect light to below deck, here they refer to shutters being fixed in preparation for a storm, the probable cause of the sailor's awakening.

George Henry Boker's *The Shark: a ballad*, from his collection *The Lesson of Life* (1848), plays to long-familiar stereotypes: the shark as the sailor's ultimate enemy, a scavenger of the sea that can instinctively detect vulnerability. Here, he describes the greedy shark as following the ship, in which a man lies gravely ill, like a vulture following a hearse. The sick man's enemy, 'Black Edward', torments the invalid by telling him, 'A shark swims under the lee!' The sailors bait a hook to catch the shark but the shark resists it, knowing that there is a sick man aboard. Once the patient dies, the corpse is bagged and thrown overboard, to the horror of the other sailors who look on as the shark tears the bag open and grabs the body.

> 'Don't throw him,' they cried, 'to yonder shark!'
> Black Ned kicked over the plank.

With a bound sprang forth the greedy shark,
And his horrid jaws upraised,
He tore the shroud like a rotten rag—
The crew in pale horror gazed.

He dragged the torn body from the shroud,
And the sea was red with gore—
'God shield us!' they cried, and closed their eyes;
But the shark was ne'er seen more.

(43–52)

However horrific this scene may appear, there is some consolation that the shark is only interested in dead men. The reference to the shroud being torn like a rotten rag reinforces this idea. Once the shark has done its job of clearing up the dead body, it is not seen again. The horror of the crew looking on is therefore intriguing. Since the corpse cannot be kept on board without becoming a site of further stink and infection, throwing it overboard is necessary. The lack of emotion with which Black Ned feeds his mate to the eager shark is dramatically juxtaposed with the sailors' sentimentality over the dead body. In the moment that the body goes overboard, they each imagine a similar fate for themselves, not as dead – since they would not feel a thing – but as eaten alive. The poem establishes a clear delineation between the pragmatism with which a dead body is dispatched and the poetical drama of the emotion that accompanies it.

Hardwicke Drummond Rawnsley is perhaps better known as one of Britain's greatest conservationists of the Victorian period. In 1895 he, along with Octavia Hill and Sir Robert Hunter, founded The National Trust for Places of Historic Interest or Natural Beauty, an institution which survives to this day with the support of more than four million members (Ranlett, 1983). Rawnsley also had a passion for bravery, which he celebrated as part of the national heritage in his collection *Ballads of Brave Deeds* (1896), to which G.F. Watts wrote a stirring preface. Watts promised, 'When our material successes may be more than rivalled by peoples and states now undreamed of; inspired by the highest human impulse, by the secret of these deeds, the beating of the great heart of the nation will vibrate through all time' (Rawnsley, 1896:viii). It is not surprising, therefore, that Rawnsley felt a poem was called for following a shark attack incident in Newcastle, Australia, on 28 November 1894. The poem is entitled *A Modern Viking: Newcastle, Western Australia* (1896) and draws on its ancestral roots to explain the fortitude displayed in the face of a shark that had seized upon the arm of a swimmer. The great Leviathan of the Book of Job is interpreted as the shark, assigning it a mythical status. The poem recounts how the shark's 'cruel fin' emerges undetected among the swimmers till the shout of 'shark', which subsequently gained its own 'iconic' auditory status, reaches their ears, 'And every swimmer felt the land was far.' The fear of being attacked by the shark suddenly creates an illusion of distance from the shore, turning it into a subjective, rather than objective, space. As they all flee for their lives there is one call to stay from the victim. His brother hears him and turns back to rescue him:

Swam ever 'twixt the shore and death,
A fence of life against Leviathan,
Felt at his back the monster's breath,
But only thought him of the wounded man.

Then, while the waves for joy broke loud,
He pushed his fainting brother safe ashore,
Nor knew a whole wide world was proud
Of him who dared, as Sea-kings dared of yore.

(29–36)

Here the shark is the 'monster', who deprived of his prize pursues the hero. The idea of the shark 'at his back' serves here only to exaggerate the courage and strength of the rescuer. The breaking waves are personified as cheers of joy while 'a whole wide world' celebrates the man's bravery. Although returning to the shark to rescue his brother is no doubt a courageous thing to do, when read from the shark's perspective the 'whole wide world' is against it for test-biting something in its territory. The ocean is personified so that its waves can convey 'joy' at the rescue of the boy from the shark. Ultimately, the poem is about the triumph of man over the shark on a mythical level, the 'Sea kings' of old against the great Leviathan. It pits men and sharks as universal enemies in which men are good and sharks evil, and deliberately sensationalises the threat of sharks for the purposes of extolling the power of men.

In another poem by Rawnsley, *M'Dermott's Deed* (1896) based on another true story, a Boatswain, McDermot, is anchored off Uzi island, south of Zanzibar, while several of his mates go for a swim. When he sees a large shark approaching, he immediately jumps on top of the shark, scaring it off and allowing time for the men to climb aboard to safety. McDermot is subsequently awarded a silver medal by the Royal Humane Society. Divers to this day recount comparable experiences of falling onto sharks, even great white sharks, and scaring them into a hasty departure. In the poem the shark is described as having a 'nostril keen' and 'eager eye', which could generously be credited as accurate observation. This is one of the few poems of the period that actually identifies the great white shark explicitly, though the accurate details are accompanied by the usual exaggerations of its mouth:

His teeth are cruel as a saw,
His throat as hell is wide,
Yea, cavernous as Death his maw,
As Death, unsatisfied.

(5–8)

So much inaccuracy in such a short passage! The popular perception of shark's teeth is that they are razor-sharp and saw-like. In fact, the teeth even of the great white, tiger and bull sharks at the top of the Shark Attack File are fairly blunt with slight serration, no worse than a regular dinner knife. What makes them effective is the considerable bite force that the jaws exert on soft-tissued mammals (Ferrara et al., 2011). In this poem, the assumption is that this is a male shark, though gender would not be something distinguishable from the surface of the ocean. The references to 'cruel', 'hell' and 'Death' are all representations of the shark as wholly evil. Milton's 'Death' has a lot to answer for. Here the dorsal fin is described as 'a ghoulish sail' that follows the scent of the dying (Figure 3.2). Having set up the image of an evil adversary, the poem goes on to describe the beautiful island, the spiced air and the joy of the English sailors as they swim.

> But murder watched from out the dark,
> With eyes that never sleep,
> And swift and silent lo! a shark
> Came upward thro' the deep.
>
> The ocean-hound athirst for blood
> Sped straight toward the shore,
> A fin was seen above the flood,
> Where was no fin before!
>
> (57–64)

Rawnsley's shark is a murderer in waiting. His reference to the shark as having 'eyes that never sleep' is made in order to suggest that like death itself it can strike at any moment. It certainly contributes to the suspense in the poem, but it is again based on a misconception about sharks. While sharks do oxygenate by allowing water to pass through their gills (ram ventilation), they can still rest in an area that provides a constant stream of water flowing over their gills. Some pelagic species such as great whites must keep moving (are obligate ram ventilators) and therefore are not able to 'lie in wait'. Other species of sharks can remain motionless on the bottom and move water over their gills by opening and closing their mouths (buccal ventilators) and by moving large spiracles situated behind their eyes (Wcisel, 2012). At present, it is unknown whether sharks sleep. A second misconception that this poem raises is the ever-open eyes. Although sharks do not have eyelids in the way that we understand them, most of them have translucent nictitating membranes that can rapidly cover the eyes to protect them, for instance just before impact with an object in the ocean or during feeding (Figure 3.3). So just as sharks don't need noses to breathe, they don't need eyelids to rest, but they have versions of them that are adapted for marine environments. Great white sharks simply roll their eyes into their heads for increased protection. The poem describes the shark ascending suddenly through the ocean. Indeed, great white sharks can speed to the surface at an estimated 56 kph (Martin, 2017). While this speed is sufficient to stun potential prey, it also means that once the shark is on its course it cannot easily stop to reassess its target – probably the cause of some collisions with humans. The poet uses absence and presence in the same moment to convey the surprise appearance of the fin, 'Where was no fin before'. But the poet's use of the words 'fierce fin' indicates how the shark becomes attached to human emotions through cultural misrepresentation.

> The fierce fin sank from out of sight;
> And e'er M'Dermott cried,
> The murderous shadow turned to white,
> The shark was on its side.
>
> No time had he to doff his coat,
> Who scarce had time for dread;
> The Bo'sun sprang from out the boat
> Full at the monster's head.
>
> (65–72)

The description of the fin disappearing beneath the waves before a strike is accurate, as is the roll onto the side to take a bite test, revealing its white underside, but in most cases the shark would then swim away. Before it has a chance, McDermott jumps onto the shark's head, and it flees. The final stanzas of the poem celebrate his great heroism in leaping 'unarmed to face and fight/The fierce leviathan' (83–84). The several mentions of sharks in poetry of this period as Leviathans suggest that this was indeed the accepted interpretation of Job 41. While the poem provides a fascinating and largely accurate account of some great white shark behaviour, all motives and emotions are supplied by the poet. Great white sharks do not race around the oceans to satiate their thirst for human blood, neither can they be described as monstrous in terms of their size, in an ocean of whales.

The Gallant Teuton (1887) by H. Hartwell, about a shipwreck off the coast of Africa in 1881, was probably written around that time, and was subsequently published in a miscellany. As those on-board sink beneath the waves, the shark is described as excitedly rushing to meet them.

> And the ravenous shark from his cave of gloom
> Hurries forth through the dark ocean's depth,
> And frolicking round their wave-hidden tomb,
> He gloats o'er the havoc of death.
>
> (13–16)

The adjective 'ravenous' seems to be permanently attached to sharks at this stage. Sharks are being described as eating hungrily, greedily, ravenously, which is contrary to what we know about all species of sharks, including the larger predators. The shark in this poem seems to exist only for moments such as these, preying on the misfortune of others. The shark is only drawn from his gloomy, dark sea world by the prospect of an exciting shipwreck meal. 'Frolicking' is something that great white sharks have not been observed engaging in, but the poet uses this playfully happy shark image to make the impending doom of those on the storm-wrecked ship, who will probably drown, more grotesque by contrast. It reinforces the image of sharks as enemies to man. Man's destruction is the shark's joy, however appallingly crude the depiction, or for that matter the rhyme.

James Russell Lowell captures the restlessness of life at sea and the longing for land in his poem *The Sirens* (1890). Lowell travelled often, but one senses from his *Fireside Travels* (1864) that the sea was not his favourite spot. He begins his chapter on his sea voyages with the observation that 'the sea was meant to be looked at from shore' (Lowell, 161:1864). Throughout his poem, the Sirens juxtapose the horrors of dying at sea with the safety of shore, but it is not called *The Sirens* for nothing. The mythical sirens of ancient Greek and Roman seafaring culture sang sailors to shore, wrecking their ships on hidden rocks – not hidden sharks – and other dangers surely await sailors who are lured by the sound of Sirens. Here the worst horror is described as that of being eaten by sharks:

> Look down beneath thy wave-worn bark,
> Lean over the side and see
> The leaden eye of the sidelong shark
> Upturnëd patiently,
> Ever waiting there for thee[.]
>
> (60–64)

The poem is indicative of Lowell's darkest hours at sea as he contemplates the leaden eyes of the shark, which suggest both drowning and the likelihood of being eaten soon thereafter. However, his journal has its lighter moments, for instance when he considers Noah, stuck on his Ark, and then observes with some envy that he was probably the last navigator to receive no bad news from home (164).

In Sir Edwin Arnold's *The Voyage of Ithobal* (1901) we find one of the more graphic poetic descriptions of a struggle with a shark. An idyllic scene of girls bathing in a creek is established, until Nesta, who is waiting in the boat, notices the 'glitter of grey light beneath the green' (1926) and realises that there is a shark between the girls and the boat. As the shark surfaces, its fin is clearly discernible. Nesta calls out that there is a shark in their midst. The simple switch from an indefinite to an emphatic definite article defines the shark as the deadly obstacle in their path back to the boat. The shark now becomes 'the fierce murderer of the deep' (1930), suggesting, yet again, that there can only be one outcome to this encounter.

> A black fin cuts the water! Nesta cries:—
> 'A shark! the shark!' and then her countenance
> I first saw fall; for, 'twixt the maids and ship
> Steered the fierce murderer of the deep, aware
> Of his sure prey; and they, aware of him,
> Bent anguished eyes on their pale mistress there.
>
> (1927–1932)

The poem switches between Nesta, the shark and the girls in the water to increase the dramatic tension of the moment. The iconic shout that conjures up the image of the great white is heard before the poet suggests that either the girls must die or their rescuers will be sacrificed. Suddenly, two African slaves come to the rescue quite spectacularly:

> A lance-head lay on deck, barbed at the point,
> The shaft new sharpened for its ashen pole
> A cubit long. Gondah strips off his gown,
> Grips the sharp steel, and rolls the cloth around,
> Leaping into the sea; so Handah too
> Holding his fighting-knife. With this the boy
> Strikes at th' attacking fish, who hath in front
> Young Gondah swimming. Savaged with the stroke,
> The monster turns to seize; opes his fell jaws,
> Toothed terrible, forgetting what he sought–
> Those naked maidens. Look! the fearless boy
> 'Tween jaw and palate of that dreadful beast
> Thrusts the wrapped spike. The murderer closes down
> The cruel mouth, but hath a bridle fixed
> Will ride him to his death. Mad wallows he
> While Handah stabs and stabs. All impotent
> Rolls the baulked fish into the crimsoned depths[.]
>
> (1939–1955)

This poem might appear to be about great heroism, but at no point is there any indication that the shark is actually attacking the girls. It is merely swimming with the girls when Handah and Gondah begin to attack it. Gondah grabs a lance and leaps onto the shark while Handah stabs it till it sinks, bloody and 'impotent', its jaws clasped together with the makeshift hook. It is not until the shark is attacked with a knife that there is any sighting of its jaws. Sticking an arm into a shark's mouth is unlikely to leave Gondah unscathed, but this vicious attack on the shark is dressed as an heroic act of love. The words used to describe the shark are 'fin', 'attacking fish', 'monster', 'fell jaws', 'Toothed terrible', 'jaw and palate', 'dreadful beast', 'murderer' and 'Mad', until 'All impotent/Rolls the baulked fish into the crimsoned depths'. It only becomes a harmless fish once it is dead, which seems rather harsh treatment of a shark that just happened to be in the water when some maidens decided to skinny-dip.

Humorous shark attack poems

Although much of the poetry that deals with shark attacks is gruesome and generally treated as tragedy, there is a body of poetry that reminds us that humour is sometimes more effective. As the songsmith, actor and composer of that Last Night of the Proms staple 'Tom Bowling', Charles Dibdin knew that it was always good business to include a shark among the perils that faced brave 'tars' (sailors). In *Ben Backstay* (1807), Dibdin relates the stories of several sailors who were all killed at sea: Spanking Jack is washed off the sail by a spray, Whiffling Tom has his head shot off and Bonny Ben Backstay is a boatswain who accidentally gets too drunk on a double dose of grog and falls overboard into the path of a waiting shark. After each stanza the refrain warns that grieving is folly and this is all part of life at sea: the shark is anthropomorphised as a worthy match for these rough and ready boys, lending an element of fairness to the scenario:

> A shark was on the starboard – sharks don't for manners stand –
> But grapple all that they come near, just like your sharks on land;
> We threw out Ben some tackling, of saving him in hopes,
> But the shark had bit his head off, so he could not see the rope.
> With a chip, &c.
>
> (11–15)

These shanty songs were supposed to help the sailors keep up their spirits during laborious work, but they also kept them in rhythm, which could be especially useful when handling decking rope and other large pieces of ships' tackle that required many hands on deck. Here the poem sets up the shark as a creature that cannot be argued with, like the land sharks, thereby creating the impression that Backstay is doomed (Figure 3.4). Even so, his loyal mates throw him a line to save him. The humorous punchline offers an incongruous image of Backstay trying to find the rope without his head. The moral of the poem is also ludicrous, as Backstay's ghost warns other sailors that it was his fault that he lost his head to the shark, so in future, 'Why, never mix your liquor, lads, but always drink it neat' (19). Crucially, the poem does not say that young Backstay was eaten by the shark, just beheaded, which seems a fitting conclusion for a sailor who had already lost his head to grog.

In Thomas Hood's *Sally Simpkin's Lament or John Jones's Kit-cat-astrophe* (posthumously published in 1906), Sally is depicted taking receipt of a portrait of her husband painted to the waist. The ghost of her husband, Jones, explains why:

> Oh! Sally, sharks do things by halves,
> Yet most completely do!
> A bite in one place seems enough,
> But I've been bit in two.
>
> You know I once was all your own,
> But now a shark must share!
> But let that pass – for now to you
> I'm neither here nor there.
>
> (13–20)

He observes the speed with which the shark bit off his legs and then humorously describes how he is now a divided person, just as his story must break off – with his portrait lacking his blue trousers.

> One half is here, the other half
> Is near Columbia placed;
> Oh! Sally, I have got the whole
> Atlantic for my waist.
>
> But now, adieu – a long adieu!
> I've solved death's awful riddle,
> And would say more, but I am doomed
> To break off in the middle.'

(33–40)

Michael Rossetti noted of Hood that he 'appears, by natural bent and permanent habit of mind, to have seen and sought for ludicrousness under all conditions – it was the first thing that struck him as a matter of intellectual perception or choice' (Rossetti, 2014). Elsewhere we have witnessed the horror and fear with which sharks are treated in poetry through the ages. The juxtaposition of a horrific attack with the complete lack of emotion from the couple creates an incongruity that serves the other implications of being separated humorously.

In Lynda Howard Sigourney's poem *The Milliners and Fishes* (1841) a shark sidles up to a boat, instilling fear in those aboard. They speed up to get away from the shark, but on closer inspection the shark is merely asking for help to be released from a piece of clothing it has been caught up in.

> A Shark in a small boat's wake follow'd long;
> The sailors thought his purpose was to eat them,
> And spread all sail; but 'twas to be unlaced,
> For he a pair of corsets had rigg'd on,
> With busk and bones, just fashionably tight,
> But could not bear the torture; so with haste
> Panting and flouncing, sought to be released.
> Item: Would it not be the surest way
> To kill that hardy and voracious fish,
> Which ofttimes foils the harpoon?

(46–55)

The incongruity of a shark in a corset makes this poem amusing, of course, but how the shark came to be wearing a corset is not explained. The reader is left to guess whether perhaps the clothing is the remains of a meal. The humorous suggestion is that although sharks often escape harpoons, corsets are nearly inescapable – as fashionable ladies of the day would have known. However, there is a more serious side to this issue. Many who dive with sharks regularly can testify to having had to save sharks from some human hazard or other, whether it be hooks, nets, lines, plastics or other human waste (Colmonero et al., 2017).

In another humorous representation of the shark, in a popular sea shanty called 'Rhyme of the Chivalrous Shark' (1904) by Wallace Irwin, the shark is described as eating almost everything but with two exceptions. It begins:

> Most chivalrous fish of the ocean
> To ladies forbearing and mild,
> Though his record be dark, is the man-eating shark,
> Who will eat neither woman nor child.
>
> (1–4)

The shanty continues with an instance of a woman who falls into the sea, only to be proffered a fin so that she can be restored to the boat. The 1904 volume was prefaced with an illustration by Peter Newell of a great white shark wearing a top hat offering his services to a drowning damsel, with a caption taken from the poem: 'He bowed in a manner most polished,/Thus soothing her impulses wild' (19–20). Having met the lady's immediate needs, the shark then turns with just as much composure to the sailors:

> So they took her aboard in a jiffy,
> And the shark stood attention the while,
> Then he raised on his flipper and ate up the skipper
> And went on his way with a smile.
>
> (33–36)

The shanty is written in the style of an Edward Lear nonsense poem or limerick. It uses humour to make us laugh at our fear of sharks and question the designation of the shark as 'man-eating'.

In Rick Noguchi's *Kenji Takezo Fights Back* (1996) the rather humorous relationship depicted is between a surfer and a shark. When a shark gets in the way of a great wave, the surfer bites the shark's dorsal fin, humorously alluding to the many large test bites that sharks have taken out of surfboards, except, of course, this one is fuelled by revenge.

> Now, he is an embarrassment.
> The other sharks will see
> How he failed to uphold his duty
> The image of the shark as terror.
>
> (11–14)

The shark's powerful identity, its 'sharkhood', is taken from it when its fin is damaged. This is the first poem in the language that personifies the shark in this way, although cartoons frequently lampoon human foibles by transposing them to sharks. His shark manhood has been stripped and he imagines how the other sharks will respond to this humiliating iconoclastic 'de-finning'. The image of seals laughing at his wound transforms the 'lethal' shark into a pitiable source of fun. Forever damaged in this shark attack, it swims away 'with its fin down' (44).

As an ardent surfer himself, Noguchi is well aware of the nature of surfing a dangerous environment as well as surfers' stories of close encounters and skirmishes with sharks. The downward pointing fins are usually a sign either of imminent attack or an imminent change of direction, but in this poem the fins are a display of humiliation. Human body language used in the anthropomorphisation of the shark allows the poet to convey an emotional side to a creature that is often depicted as lacking any emotion at all. Amusingly, we are asked to sympathise with the shark, attacked by the vicious surfer. Nevertheless, the poem raises an important issue of the vast scale of human attacks on sharks, as a revenge response to the minuscule number of shark attacks on humans, and as part of a world-wide business in shark fins and flesh.

Sharks in other cultures

Here we turn from the comic to more disturbing poetry. John Leyden wrote a set of poems based on his experience on the Island of Sagur (1819), including his witnessing of infant sacrifices. He relates how the custom was to offer infants 'to glut the shark and crocodile', to fend off evil powers. In this poem the mother looks on the playful smile of her child as she tosses it into the Ganges:

> To glut the shark and crocodile
> A mother brought her infant here:
> She saw its tender playful smile,
> She shed not one maternal tear;—
> She threw it on a watery bier:—
> With grinding teeth sea monsters tore
> The smiling infant which she bore:—
> She shrunk not once its cries to hear!
>
> (9–16)

Although we are told that the Ganges is a place where 'man-devouring monsters swarm' (32), the true horror in the poem comes not from observation of the shark or the crocodile, here incorporated as grinding and tearing monsters, but from the mother who shows no emotion as her child is eaten. The poem ends with a reflection that sailors passing by will remember it as a place 'Where Mercy ever sued in vain' (48).

In *The Maori's Choice* (1866) by C.P.H. the Maori is asked about his diet. The poem begins epigrammatically with *Chacun a son Gout* (Each to his own taste). A suggested list is made, to which the Maori agrees – yes, it is all part of his diet, including 'shark when putrifying' (12), but he tells the enquirer he has left off the best meat – man. What is most interesting here is the reference to the bright shiny teeth of the shark-eating Maori.

The rascal's teeth were sharp and bright,
He smack'd his lips with relish;
No thought had he, the savage wight,
Of harm in feasts so hellish.

(21–24)

The Maori cannibal is made to bear some of the characteristics of the shark. Aside from his sharp white teeth, he is apparently indifferent to what he eats:

He fasts, or feasts, just how he can;
The rest is visionary;
But best is pleased with flesh of man,
Served à la Missionary!

(33–36)

The final line is intended as a joke in bad taste, but it is also a sharp reminder of what could actually happen to missionaries in their quest to bring Christianity to Pacific islands. After the Treaty of Waitangi (1840), stories of Maori cannibalism subsided and Fiji became a focus of attention (Berker, Hulme et al., 1998). The Maoris have long incorporated sharks into their own poetry and folklore. One Maori proverb, '*Kaua e mate wheke mate ururoa*' (idiomatically translated as 'Don't die like an octopus, die like a hammerhead shark'), is a prompt to always show the tenacity and fight that a hammerhead shows when caught by fishermen (Woodward Maori, 2017).

James K. Baxter is widely regarded as one of New Zealand's most successful poets. An unconventional character in many ways, he showed from an early age great interest in, and concern for, New Zealand's natural landscape and its indigenous people (EO'R, *Literature online*, 2008). His poem *The Maori Boy* (1980) explains the process by which a shark is hunted, beaten to death and baked in a makeshift fire. To the Maori boy a shark is simply a meal for a day. The boy will then use the same fire to sleep beside until the next day. The poem is iconoclastic, refusing any of the cultural representations by which sharks are mis-recognised, situating it within a context in which the boy and the shark hunt merely to exist. However, in another poem the appropriation of sea life for ornament is condemned. Louis Aston Marantz Simpson is a Jamaican-American poet. In *Moving the Walls* (1988) the poet reminds us of the many uses that humanity has made of sharks. He relates the adventures of the Prince of Monaco with the ensuing trophies. He has a walking stick made from shark backbone; though sharks are all cartilaginous fish – that is, they have no bones in their bodies – such sticks are among an array of marine creature adornments.

The poet suggests that the magnificent marine creatures are reduced to the banal once they are ripped from their environments to become lifeless and colourless dead artefacts in domestic contexts. The killing of aquatic life to serve the needs of humans is representative of the selfishness of humanity more generally.

> For Leviathan does not exist,
> And the sea is no mystery.
> For a shark is a walking stick.
>
> (72–74)

The poet points out that after centuries of exploration, the wonders of the sea are repeatedly robbed of their mystery to decorate human environments: nature is reduced to human dimensions to the detriment of all when 'a shark is a walking stick'.

The rock-and-roll subculture could also be considered foreign. Edward Field's *Shark Island* (1992) is the first poem in the English language that discusses shark repellent. The tale is somewhat reminiscent of an episode alluded to in 'Mud Shark' (Zappa, 1971). The song refers to an actual explicit event in a hotel room in 1969 between members of the pop group, a woman and a dead fish. Field's poem begins with Biff Harmon and his wife Wilma. Biff is a shark behaviour expert who keeps sharks in his lab, and swims with 'killers' (3). He decides to take a trip to Central America to study an undocumented freshwater shark. We are told early on in the poem that the natives of this area still make human sacrifices to pacify the shark spirits. We are then told that the natives have had a run of bad luck, crop failure and starvation, and that the pigs have already been fed to the sharks. They are looking for a change of fortune when plump Wilma turns up. Wilma and Biff take a boat out to see the sunset and in the middle of their love-making a shark takes a bite out of Wilma's ecstatic overboard leg.

Biff manages to rescue Wilma by hitting the small shark with an oar, but the poet suggests that now the sharks will stop at nothing to get her:

> Wherever she went, even on land,
> they tracked her from the lake with their radar
>
> (52–53)

As we know, sharks do not go crazy for human blood, but the suggestion is that now they have tasted Wilma they will stalk her, in an interesting twist on the shark/land-shark relationship we have seen in earlier representations, until she steps into the lake again.

Wilma is finally snatched, drugged and prepared as a shark sacrifice in a horrific and sexually explicit episode that aligns more closely with film representation than the poetic tradition. The shark is invited to nose her menstrual blood from its source with the idea that the Great Tumba shark will be satisfied by her and bring better fortune to the village. Mungo, the Shaman's son, endeavours to save Wilma from this fate with the help of a shark repellent:

> Here Mungo became invaluable with his native lore.
> He had a vial of extract of the rare shark orchid,
> that, spilled into the water, would repel the sharks awhile.
> It was a trick his father used
> to make people think he had power over sharks.
>
> (79–83)

The outcome of the research field trip to investigate this rare shark is that the villagers all get eaten and the visitors return home. The representations throughout of Wilma, the villagers and the sharks are all deserving of scrutiny, but our focus here is on the sharks. The sharks are represented as drug addicts, God-like spirits with supernatural power over the fortunes of men. They are crazy for human blood and they are very unhappy about being offered white meat and then having it snatched away. When Mungo, the faithful servant, wrestles Tumba and ends up being eaten it is only noted that he hasn't time to wave as Biff and Wilma reach the shore and head home. The poem reinforces the message that has been a staple of New Historicist and postcolonial literature for two decades, and more recently for eco-criticism, that so-called 'civilized' society can have devastating consequences for native cultures when they collide (Greenblatt, 1988). The poem is loaded with such vulgar misrepresentations, including the disturbing supposition that sharks are more interested in human menstrual blood. Although the presentation of sharks as gods is presented as dangerously deranged, yet there is more foundation for this than even the poem allows. We all depend on sharks to keep the oceans clean, after all.

Shark fishing

Sharks are extensively overfished across the world's oceans. The IUCN (International Union for the Conservation of Nature) has raised the alarm on this issue, and it remains hotly debated how long we can maintain present levels of shark fishing if we expect to continue to live on a planet that requires a healthy ocean. Poets have been engaging with this topic as far back as the mid-nineteenth century. A more sympathetic approach to a shark caught in a net is *Shark Meat* (1970), in which Gary Snyder describes his reflections upon a lunch with native islanders.

This poem is the first to consider a shark encounter from the perspective of a dead shark. A chance encounter with a net sees the beginning of the shark's transformation, which ends as a community meal: 'The island eats shark meat at noon' (14). However, the majestic shark's ignominious ending is redeemed through reflection and reverence, as Snyder's biographer, Bert Almon has argued: 'Poems such as "Song of the Taste" and "Shark Meat" celebrate eating as a kind of sacred act' (Snyder, 1979:37).

Dave Smith is a prolific American poet, described by Mark Johnson as drawing 'strength from his place and time and people, but [who] expects us to read his poems and connect to more universal truths, beauties, and goodnesses' (Johnson, 2005:93). This is perhaps not an obvious introduction to *The Shark in the Rafters* (1974), in which Smith describes in graphic detail the slaughtering of a shark, following an attack. The shark is dissected and distributed piece by piece to the starving villagers, until all that remains is the jaw.

In an interview with Suarez, Smith commented that he regarded himself as a 'regionalist': 'My sense of appreciation of what life means or could mean, whatever I know about life, stems from a sense of place, a sense of the ghostliness of meaning, in that place where the apprehension of meaning was attached to visible realities: an ocean for me always represented something. . . . when I'm not near an ocean, I don't feel complete' (Suarez, 1999:28). It is all the more interesting that he should depict the shark stripped of everything, from its last meal right back to its mythical jaw – the single marker of its identity that still lends it integrity.

Greg Pape also draws extensively in his poetry on the impact of nature on emotion. In an interview with Drew Pomeroy, he argued:

> If our poems and other works of art help us live our lives, and that seems to be one of the primary purposes of art, it is by articulating, questioning, and shaping experience, sometimes making sense, providing insights or feelings that can be shared, sometimes just putting something out there we don't completely understand, adding to the conversation. If we have learned anything it's that the human element is not something apart from Nature but something within Nature.
>
> (Pape and Pomeroy, 2013: online)

In *Sharks, Caloosahatchee River* (1984), Pape begins by establishing 1950s culture. Cars commonly had fins. Here, those 'fins' are lined up. This is not incidental but part of their owners' identities, for these are shark fishermen.

> He watches as men line up on the dock
> the bloody sharks they caught.
> The fins gleam like the waxed fins
> of the new cars. Their eyes don't close either.
>
> (21–24)

As the men line up on the dock, the boy is able to see their catch. The sharks are bloody, suggesting a mortal struggle. Having scanned the sharks, a boy scans the bodies of the cars till he notices the headlamp and, reinforcing the myth that sharks don't close their eyes, emphasises the connection between the cars and sharks of these men. Once the men have taken their snapshots of their catch they throw them back, but the sharks are as dead as the cars. The lives of the sharks are now just stories to be told about a sport fishing trip to Florida.

The conflation of environments, the fins and the qualities of silver, links the images into reflections that become the boy's own reflections on the interconnectedness of objects and events. He is fully absorbed into a space of fictional possibilities. The poet imagines the shark body drifting to the riverbed to join all of the other victims. As the boy later watches a matinée, he reflects that the silver screen is like the river and the cars with their fin-like wings that 'slice the sunlight'. Later the boy drives one of the cars into the river, steps out and watches it with everyone else as it sinks like a shark. There is then a sudden turn in the poem as the poet considers that the boy may not have escaped and that he, too, has drowned. A search for his body is in vain as he becomes one of the riverbed sharks.

With the disappearance of the boy, the poem veers from surrealism toward myth. We see the boy, at first an observer himself, enter into the story initially via the car fins before he witnesses the shark fins and becomes the subject of the poem as shark boy. What is clear from the poem is that these shark fishermen have no regard for the sharks: they are phallic identifiers, acquired to establish a hierarchy among them, like the finned cars that the men drive.

By way of contrast, James Fenton's *The Milkfish Gatherers* (1993) describes the catching of a shark as a hard-won prize for a poor community of fishermen. The excitement as the shark is offloaded from the boat is discernible in the waiting 'crowd' that has gathered on the beach. The gills are cooked, the liver is cut out and, in a very arresting image, the shark's teeth appear to be sharks 'advancing along a jaw'. These teeth make the shark seem embattled, with reinforcements at the ready. In fact, sharks replace their teeth throughout their lives, since the hard bones of their prey frequently break them. In this poem the shark is dissected and shared among the community that is desperate for provisions. The massive liver referred to here is characteristic of sharks and helps them to maintain their buoyancy because it is filled with oil – which is lighter than salt water. Taken out of the shark, the liver 'wallows' on the shore, suggesting an unwieldiness that is indicative of the shark's nature, impossible to contain. What is interesting about Fenton's poem is the context of gratitude that reminds us of the complexity of the issues surrounding shark fishing.

We end with Andrew Motion's *Sailing to Italy* (1997). Here the poet is preparing for a journey and is reminded of John Keats coughing up blood and heading to Rome to improve his health. He gradually extends this idea into a broad reflection on the sea journey and Italy itself. Motion is well known for his interest in water, of both the salt and the fresh variety, that is able to traverse the range of natural and artificial environments, and he is able to harness it for the creation of a wide range of poetical effects. In this poem, during one extended narrative while on the ship, he reflects on the cultural impact of the ocean on civilization. The shark is included as an afterthought. Hooking and hauling on deck would not bring out the best side of most creatures, but here there is no sign of remorse for mistreating the shark, only a description of its 'bitter eyes' as 'hatred pooled in slits of oily dark' (344).

The shark's denticulate skin is denatured and transformed into a coat of armour, but this battle has been lost to the metal hook that has also dislocated the shark's, interestingly vulnerable, jaw. What is so shocking about this representation is the absence of emotion on the part of the protagonist, who merely takes a photograph as some sort of proof. Writing about his role as Poet Laureate (1999–2009), Motion argued that poetry 'stretches our humanity – not in the sense that it tells the truth to power and deepens our knowledge of the human condition, but in the sense that it can have a liberalising effect upon our natures' (Motion, 2000:87). Interestingly, he resisted the word 'liberating' for 'liberalising', suggesting that poetry frees less than it changes states or states of being.

While there are yet many more ways in which sharks could be represented poetically, perhaps the general decline in interest by poets is rather a manifestation of a more sensitive, or environmentally friendly, approach to these creatures. Poets wield great power to change minds, and their contributions to a re-presentation of sharks in positive ways are to be welcomed.

References

Primary works are derived from Chadwyck-Healey Literature Collections online.

Almon, B. (1979) *Gary Snyder.* Boise, Idaho: Boise State University.

Barker, F., Hulme, P. and Iversen, M. (1998) *Cannibalism and the Colonial World.* Cambridge: Cambridge University Press.

Beckert, M., Flammang, B.E. and Nadler, J.H. (2015) Remora Fish Suction Pad Attachment is Enhanced by Spinule Friction. *Journal of Experimental Biology.* 218, 3551–3558.

Brickley, R. and Wagner-Martin, L. (2016) *Understanding Sharon Olds.* Columbia: The University of South Carolina Press.

Colmanero, A., Barria, C., Broglio, E., Garcia-Barcelona, S. (2017) Plastic Debris Straps on Threatened Blue Shark *Prionace Glauca. Marine Pollution Bulletin.* Vol.115.1–2, 436–438.

Domeier, M.L. (ed.) (2012) *Global Perspectives on the Biology and Life History of the White Shark.* Boca Raton: CRC Press.

Dowson, J. (2016) *Carol Ann Duffy: A Poet For Our Times.* London: Palgrave Macmillan.

Ferrara, T.L., Clausen, P., Huber, D.R., McHenry, C.R., Peddemours, V., Wroe, S. (2011) Mechanics of biting in great white and sandtiger sharks. *Journal of Biomechanics*, 44: 430–435.

Fraser, G. (1961) The Poetry of Thom Gunn. *Critical Quarterly.* 3.4, 359–367.

Gallagher, N. (2011) 'Partial to Some One Side': The Advice-to-a-Painter Poem as Historical Writing. *ELH*, 78.1, 79–101.

Gerstler, A. (2006) *Light and Shade: New and Selected Poems by Tom Clark.* Minneapolis: Coffee House Press.

Green, M. (2011) Death and Nonsense in the Poetry of George MacDonald's *At the Back of The North Wind* and Lewis Carroll's Alice Books. *North Wind: A Journal of George MacDonald Studies.* 30, 38–49.

Greenblatt, S. (1998) *Shakespearean Negotiations: the Circulation of Social Energy in Renaissance England.* Berkeley: University of California Press.

Hasan, S.M. (2012) Ted Hughes' Animal Poems: An Embodiment of Violence or Vitality? *Dawn.* 1.2, 174–188.

Houston, A., and Pincus, S. (2001) *A National Transformed: England After the Restoration.* Cambridge: Cambridge University Press.

Irwin, W. (1904) *Nautical Lays of a Landsman.* New York: Dodd, Mead and Company.
Johnson, M. (2005) The Dangerous Poems of Dave Smith. *Southern Literary Journal.* 38.1, 91–115.
Jones, T. (1985) The Xoc, the Sharke, and the Sea Dogs: An Historical Encounter. In: Fifth Palenque Round Table. Fields, V. M. (ed.) Monterey, California; Herlad Printers Inc.
Kermode, F. (1991) Part and Pasture. *London Review of Books*. 13.23, 17.
Lee, D. (1993) About Al Young. *Ploughshares.* 19.1, 219–224.
Lerner, E., Austin, D.C., Kizer, C. and Matthews, W. (1986) An Interview with Carolyn Kizer and William Matthews. *A Journal of Literature and Art.*12.1987, 52–64.
Lowell, J. R. (1864) *Fireside Travels.* London and Cambridge: MacMillan and Co.
Mackey, N. and Young, Al (1978) Interview with Al Young. *MELUS.* 5.4, 32–51.
Martin, A. (2017) How Fast Can A Shark Swim?. http://www.elasmo-research.org/education/topics/p_shark_speed.htm (Accessed 13/06/2017).
Martin, R.A., Rossmo, D. K. and N. Hammerschlag (2009) Hunting Patterns and Geographical Profiling of White Shark Predation. *Journal of Zoology*. 279.2, 111–118.
Montgomery. J. (1880) *The Poetical Works of James Montgomery with a Memoir: Vol. II.* Boston: Houghton, Mifflin and Company.
Morris, D. Z. (2016) Trash-Eating 'Shark' Drone Hits the Water in Rotterdam. http://fortune.com/2016/09/11/trash-eating-drone-boat/ (Accessed 11/06/2017).
Moss-Gibbons, C., Davenport, G. and McDonald, I. (eds.) (2001) *The Royal College of Physicians and its Collections: An Illustrated History.* London: James and James.
Motion, A. (2000) Poetry in Public. *RSA Journal.* 148. 5494, 86–91.
Mullan, J. (2013) Anarchy in Peterloo: Shelley's Poem Unmasked. https://www.theguardian.com/culture/2013/jul/08/anarchy-in-peterloo-shelleys-poem-unmasked (Accessed 18/06/2017).
Natanson, L. J. and Skomel, G. B. (2015) Age and Growth of the White Shark, *Carcharodon Carcharias* in the Western North Atlantic Ocean. *Marine and Freshwater Research.* 66, 387–398.
Orlich. I. A. (1993) Song of My Emerging Self: The Poetry of Andrei Codrescu. *MELUS.* 18.3, 33–40.
Pape, G., and Pomeroy, D. (2013) An Arched and Lighted Entrance: An Interview with Greg Pape. *Poetry Matters* http://readwritepoetry.blogspot.co.uk/2013/09/an-interview-with-greg-pape.html.
Passmann, D.F. and Hermann, J.R. (2005) From 'Mossy Caves' to 'Rowling Waves': William Diaper's Nereides: or, Sea-Eclogues. *The Perennial Satirist: Essays in Honour of Bernfreid Nugel*, ed. Peter E. Firchow and Hermann J. Real. Munster: LIT Verlag, 29–47.
Percy, T. (1812) *Reliques of Ancient English Poetry Consisting of Old Heroic Ballads, Songs, and Other Pieces of Our Earlier Poets, Together with Some Few of a Later Date.* Vol. I. London: Richard Taylor and Co.
Peterson, S. (1965) *John Tutchin Selected Poems* (1685–1700). Los Angeles: The Augustan Reprint Society, University of California.
Ranlett, J. (1983) 'Checking Nature's Desecration': Late-Victorian Environmental Organization. *Victorian Studies*. 26.2, 197–222.
Rappoport, A. S. (1928; 2007) *Superstitions of Sailors.* New York: Dover Publications Inc.
Rawnsley, H. D. (1896) *Ballads of Brave Deeds.* London: J. M. Dent.
Rooney, K. (2014) About the Shark: How Do We See Ourselves in Animals? Chicago: poetryfoundation.org.
Rossetti, M. (2014) *The Poetical Works of Thomas Hood.* Adelaide: e-books@Adelaide.

Sloan, G. (1981) Profile: Even Merriam. *Language Arts*. 58.8, 957–964.
Stephens, M. (ed.) (2007) *Poetry 1900–2000: One Hundred Poets from Wales*. Cardigan: Parthian Library of Wales.
Suarez, E., Stanford III, T.W. and Verner, A. (1999) *Southbound: Interviews with Southern Poets*. Columbia: University of Missouri Press.
Timm, L.L. and Fish, F.E. (2012) A Comparative Morphological Study of Head Shape and Olfactory Cavities of Sharks Inhabiting Benthic and Coastal/Pelagic Environments. *Journal of Experimental Marine Biology and Ecology*. 414–415, 75–84.
Van Duzer, C. (2013) *Sea Monsters in Medieval and Renaissance Maps*. London: The British Library.
Walker, S.B. (2013) *The Ecological Poetics of James Dickey*. New York: Edwin Mellen.
Wcisel, M. (2012) How to White Sharks Sleep. http://www.sharkwatchsa.com/en/blog/category/482/post/1099/how-white-sharks-sleep/ (Accessed 13/06/2017).
Webster, S. (2003) *Thinking About Biology*. Cambridge: Cambridge University Press.
Weinreb, B. and Hibbert, C. (eds) (1983) *The London Encyclopedia*. London: Macmillan.
Woodward, Maori. http://www.maori.cl/Proverbs.htm (Accessed 15/06/2017).

ARTS AND SCIENCES. 169

THE SHARK.

Q. What kind of fiſh is the ſhark?

A. It is very large, and reckoned the boldeſt of all fiſh. Rondeletius ſays, that he ſaw a middling one, that weighed almoſt a thouſand pounds. Gillius ſays, that he was credibly informed, that a fiſh of this ſort, not weighing leſs than four thouſand pounds, being diſſected, had a whole man found in its belly. He further adds, that the people of Marſeilles told him, that they had caught one, in which they found a man armed with a coat of mail.

This fiſh is common in the Weſt-Indies, and is very deſtructive to thoſe that amuſe themſelves

I with

Figure 1.1 Richard Turner, *An Easy Introduction to the Arts and Sciences*, 1791 CUL 7240.d.41., used with the permission of the Cambridge University Library

Figure 1.2 Great white shark, used with the permission of Seapics

Figure 1.3 Vivienne Westbrook, *Tiger shark*, used with the permission of the artist

Figure 1.4 Bull shark, used with the permission of Seapics

Figure 1.5 Risk Taking, used with the permission of Depositphotos

Figure 1.6 Shark danger sign, used with the permission of Depositphotos

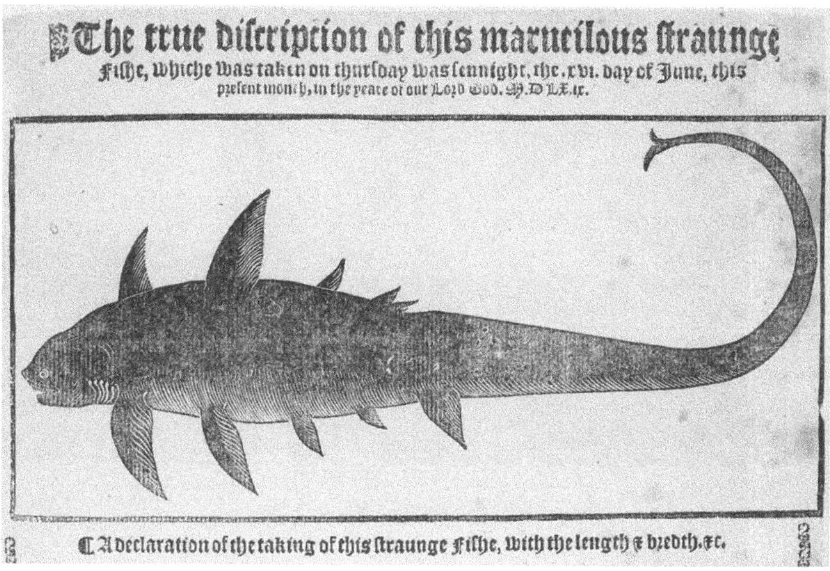

Figure 1.7 'True Discription', 1569. BL, Huth 50(41). © The British. Library Board

Figure 1.8 Jim Toomey, *Sherman's Lagoon*, used with the permission of the artist

Figure 1.9 Sharks attacked, used with the permission of Seapics

Figure 1.10 Fins for sale, used with the permission of Seapics

Figure 1.11 Cage-diving, used with the permission of Seapics

Figure 1.12 Surfboard or seal?, used with the permission of Seapics

Figure 1.13 Beware sign, used with the permission of Depositphotos

Figure 1.14 Wolfgang Leander, Smiley, used with the permission of the artist

Figure 1.15 Isabel Muscat, Wolfgang Leander dancing with a young tiger girl, used with the permission of the artist

Figure 1.16 Shark surrounded by fish, used with the permission of Depositphotos

Figure 1.17 Diver and Reef shark, used with the permission of Seapics

Figure 2.1 The fossil shark, used with the permission of Seapics (fossil) and the artist, John Long (drawing)

Figure 2.2 Shark with little friends, used with the permission of Depositphotos

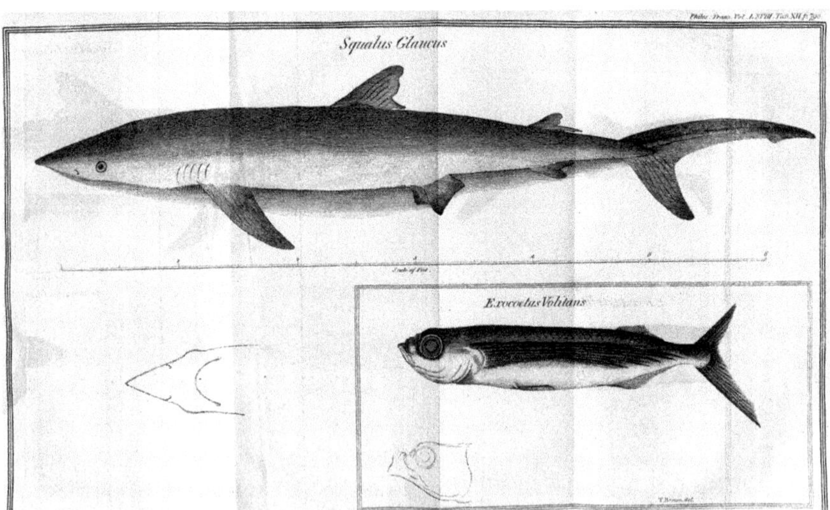

Figure 2.3 William Watson's representation of a shark, *Philosophical Transactions of the Royal Society* 68:789

Figure 3.1 Sharks as taxes, used with the permission of Depositphotos

Figure 3.2 Fin, used with the permission of Depositphotos

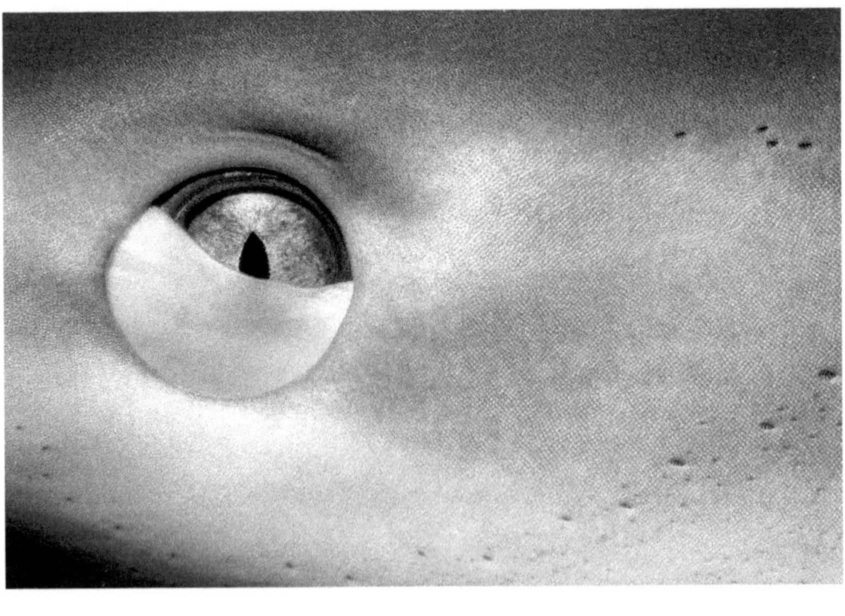

Figure 3.3 Nictitating membrane, used with the permission of Seapics

Figure 3.4 Land shark, used with the permission of Depositphotos

Figure 4.1 Sharks as pirates, used with the permission of Depositphotos

De Canecarch. & Lamia. Lib. IIII. 207

Hanc picturam Carchariæ Canis ad sceleton olim fieri curaui.

cete. de lamijs uerò priuatim agit, cum aliâs, tum libro 5. de piscatione, diuersum utrosq; capiendi modum describens, ut referemus in E. Fieri quidem potest, quod Nicander Colophonius refert, ut in nonnullis locis, carcharias alio nomine lamia dictus sit, propter similitudinem: ab alijs uerò exactiùs suo discretus nomine. Aristoteles semel duntaxat puto lamiæ meminit, cartilagineis piscibus adscribens, lib. 5. de hist. cap. 5. Canicula ferox cetacei generis, illa est quæ urinantes circa spongias graui periculo infestat, &c. Vuottonus. atqui Rondeletius hoc non ad cetaceam, sed minorem caniculam refert, quam Plinij cognominat, ut à scylio siue canicula Aristotelis discernat. Rondeletius lamiam à cane carcharia nõ distinguit: Bellonius distinguit, in hoc potissimũ, quòd lamiæ os in anteriore capitis parte tribuit, carchariæ uerò rostrum in mucronem exporrectum. *Lamiam à Cane Carch. differre.*

s 2

Figure 5.1 Conrad Gessner, *Historia Animalium* (1558), used with the permission of the Master and Fellows of St John's College, Cambridge

Figure 5.2 *The Manner of Their Fishing*, John White (1585), used with the permission of the British Museum

Figure 5.3 *Watson and the Shark*, John Singleton Copley (1778), used with the permission of the Detroit Institute of Art

Figure 5.4 *Gulf Stream*, Winslow Homer (1899), used with the permission of De Agostini Picture Library/Bridgeman Images

Figure 5.5 Jaws poster, Roger Kastel (1975) used with the permission of the artist

Figure 5.6 Polynesian Shark tooth weapon, used with the permission of the Museum of New Zealand Te Papa Tongarewa

Figure 7.1 Diver documenting shark, used with the permission of Seapics

Figure 7.2 Roger Horrocks, Wolfgang Leander at play with a curious tiger shark, used with the permission of the artist

4 Drama shark

We find the shark on the stage from the sixteenth century through to the present. It has played many parts, from that of the vicious drama critic, the serial stalker of sailors, a snatcher up of unconsidered trifles and the usual suspect for any missing bodies at sea. A vital aspect of the shark in its more tragic parts, or as pirate or predator, is that it remains unseen on stage as a discernible model for another character or by way of analogy or metaphor. In its role as sexy seducer it is even more important that the shark is discreetly embedded in a monologue. As we will see, the versatile shark is also capable of shouting out conservation issues from a fish counter.

Sharks as critics

As in the case of other writings from the Early Modern Period, the word 'shark/e' is used in the drama either as a noun, meaning a parasitical person (*OED* 1600), or a verb meaning 'to victimise' or 'to cut or tear' (*OED* V1, 1569 or V2 1596). There are fleeting references to the oceanic shark, as in Shakespeare's witches' brew, which includes the 'maw and gulf/Of the ravin'd salt-sea shark' (*Macbeth*, IV.1.23–24), but little else of note. In Early Modern Drama it is much more usual to meet with land-sharks, though there are certainly allusions to the attributes and behaviour of the oceanic sharks in the process of depicting them. For instance, in John Webster's *Appius and Virginia* (1625) there is an exchange between the clown and Virginia in which the oceanic shark is appropriated for sexual innuendo. The clown tells Virginia 'There's a certaine fish, that as the learned divulge, is call'd a sharke. Now this fish can never feede while he swims upon's belly, marry when he lies upon his back, oh he takes it at pleasure' (III.1). What is interesting is the reference to the shark as an exotic creature at this time: 'as the learned divulge, is call'd a sharke'. The clown introduces the feeding behaviour of the great white shark, which uses countershading to approach its prey, where its grey upper body camouflages its presence against the darkness of the ocean depths, but then rolls to reveal its white underbelly as it bites, or in this case, takes the bait.

In Nahum Tate's *Cuckolds-haven* (1685), based on John Marston et al., *Eastward Ho* (1605), Touchstone tells Fang that 'I will lodge my Land-shark and

Water-shark in sowce together' (III.2), but his water shark is Sir Petronell Flash, a fraudster, not an oceanic shark. In Thomas D'Urfey's *Sir Barnaby Whigg* (1681), the 'Sea-shark' refers to the Captain, rather than the fish. Generally, in the drama of this period, even sea sharks are sailors rather than real oceanic sharks. In William Congreve's *The Double Dealer* (1693), sharks are appropriated to describe critics. Here Congreve argues:

> Criticks avaunt; for you are Fish of Prey,
> And feed like Sharks, upon an Infant Play.
> Be ev'ry Monster of the Deep away;
> Let's have a fair Tryal, and a clear Sea.
>
> (Prologue)

At this time, there was a general tendency for dramatists to 'characterize themselves or their audiences as soldiers or sailors and the stage environment as some type of military situation' (Freeman, 1997:87). The critic as shark denoted an extreme antipathy that was subsequently to prove very fashionable in dramatic prologues. Congreve depicts the critic as particularly ruthless in feeding on an 'infant', new, play. Since only the shark is specifically named it must be deemed the epitome of the sea monster, falsely imagined as preying on babies. The suggestion is that only if the audience is cleared of such monsters will the play stand any chance of not being devoured. In the seventeenth century, the very idea of the seas being emptied of sharks might well have seemed unimaginable! In the process of attacking critics he represents sharks entirely negatively.

In Mary Pix's *The Spanish Wives* (1696) the prologue also registers a complaint against our fishy critics:

> For thus your Criticks serve Nine Plays in Ten,
> Worse than Jack Frenchman does our Merchant-men.
> Like Pyrates too, while honest men they're breaking,
> The damn'd Fresh-water Sharks are n't worth the taking:
> Yet long to maul these same New Plays as much
> As we, when Homeward bound, to take a Touch;
> Or, as Dubart, to snap his Brother Dutch.
> Yet why shou'd they Hiss Plays not worth regarding?
> Do we Bombard a Town not worth Bombarding?
>
> (Prologue)

While the prologue is thick with analogies of adversaries, sharks, 'not worth the taking', are particularly despised for mauling the plays, since they are the least deserving of praise themselves. Here sharks are described as hissing, clearly combining them with the mythical adversary of mankind – the snake – for an extra dose of antipathy. Yet comfort is drawn from the fact that even these critics would not bother to criticise a truly awful play. The attention of such sharky critics is thereby turned into a compliment by Pix.

Sharks in the eighteenth and nineteenth centuries

Mary Leapor has enjoyed significant reassessment in recent years, especially by feminist critics wishing to assert the importance of her reflections on eighteenth-century patriarchal culture from alternative class and environmental perspectives (Milne, 2015). In *The Unhappy Father* (1745), Leapor relates a shark attack in which the shark features again as a greedy feeder, following the lead of the many male-authored works that depict the shark in this way. Young Lord Polonius slips on a plank, falls into the sea and directly into the path of a shark. His servant leaps in after him, only to become the shark's next meal. As the Sailor relates the story to Paulus, the shark becomes the monstrous figure of 'Death' recognisable from Milton's *Paradise Lost.*

> All Help was vain; but yet his desp'rate Servant
> Leap'd in the Ocean to his Lord's Relief;
> And, as we think, serv'd to enlarge the Meal,
> And glut the Maw of that voracious Monster.
>
> (4–7)

We note that the servant is only presumed eaten. Paulus imagines the horror that Polonius' father, Dycarbus, will feel when he is told that his one great hope in life has ended as a shark snack. As Paulus relates the tale to Dycarbus, once again in imitation of Milton's depiction of Sin and Death in *Paradise Lost* Book II, the emotional chaos is palpable.

> The screaming Sailors flew to his Relief:
> But of a sudden the convulsive Ocean
> Appear'd to labour with a monstrous Birth.
> A fatal Shark, the largest of its Kind
> Roll'd his unwieldy Carcase through the Deep,
> And toss'd above the Waves his horrid Jaw.
> The stoutest Bosoms then were froze with Fear:
> But Timnus, faithful to his dying Lord,
> Rush'd in the Waters to partake his Fate.
>
> (74–82)

The shark in this tragedy takes on mythic status as the ocean breaks open to deliver a huge monster, a true contender as a model for Peter Benchley's *Jaws.* Although the great white is not named, the description offered makes it the likeliest species. As his 'horrid Jaw' breaks through the waves, instilling fear in everyone, the young servant, Timnus, jumps into the sea. In this re-telling of the tragedy, Timnus is not trying to save his Lord Polonius, but die with honour alongside his master. Upon receiving the news, Dycarbus dies. Terentia, Polonius' wife, mourns and swoons:

> Think on the Partner of my faithful Breast
> Deny'd the usual Honours of a Grave;

His trembling Flesh torn from the living Bones,
To glut the Hunger of a raging Monster.

(123–126)

For the second time in the tragedy, anxiety over the lack of a burial is raised. Polonius is imagined as travelling across the seas in the living grave of the shark's stomach. This is reflective of a widely held belief at the time that the unburied Christian soul would be lost for all eternity (Cressy, 1997). It is not merely a sentimental desire to have a spot on which to lay flowers of remembrance.

The graphic image of Polonius' flesh being torn from his bones has sufficient resonance with popular representations of religious martyrdoms to turn Polonius into one of them – sacrificed to a monstrous greed. It must be noted here that recent research has revealed that far from being greedy, sharks eat very infrequently (Marten et al., 2005). In the tragedy the 'raging' that instils a frozen fear in all who look on suggests that the shark is overwhelmingly angry, but of course this is just another example of misattribution of shark behaviour to human emotions for greater dramatic effect. Large predatory sharks tend to preserve their energy, instead employing a range of inquisitive strategies, such as circling and bumping, before they choose a target to lunge at.

As this tragedy draws to a close, all is not lost. The missing Lord and the sailor Timnus return to explain how they were saved by the fact that a dead carcass was floating by – which the shark seized instead, of course. So after all this high drama, it seems that the shark was not a raging monster after all, but a shark doing what it does best – cleaning up the ocean. Timnus relates:

Nor had we liv'd, but that the heav'nly Pow'rs
Still temper Mercy with their stern Decrees,
And sent between us and the raging Monster
The floating Carcase of a shipwreck'd Man:
His greedy Jaws devour'd the ready Prey,
And left pursuing our forbidden Lives.

(160–165)

The shark took the dead man and swam off even though Polonius and Timnus were still in its path. If it had then eaten them as well we might have agreed that it was indeed a 'greedy' shark. When we think about this episode from the refreshing perspective of the shark it seems that these humans really need to calm down. Rather than understanding the shark on its own terms, Timnus and Polonius' survival is attributed to Divine intervention. God was obviously just testing. However, the tragedy is not yet over. Terentia, ignorant of this latest news, drinks a cup of poison. Polonius arrives on the scene to bewail the deaths of his parents and wife – all as a result of a false shark-attack story. That is the tragedy.

John Delap's *The Captives* (1786) includes an epilogue by Thomas Vaughan to be spoken by Mrs Sarah Siddons, the foremost actress of her day. She assures the audience that the boat in the play has been brought safely to shore. However, the play is also a boat that must navigate a sea of critics.

> Critics, like monsters, on each side appear,
> Herald, the whale; and shark, the Gazetteer –
> If these he chance t'escape, there comes a squall
> From Lloyd's, St. James's, London, or Whitehall;
> Here Chronicle, like Scylla, guards the coast,
> There foams Charybdis – in the Morning Post.
> Mark how they break his rudder, cut his cable,
> Tear up plan, diction, sentiment, and fable;
> Their order is – an order they enjoy,
> To seize, to burn, to sink, and to destroy.
> What wonderous chance our author should survive,
> That in such boisterous seas his bark's alive?
>
> (Epilogue)

The whale and the shark here suggest two types of critics. The ruthlessness of the journalist of the *Gazetteer* is compared to the perceived ruthlessness of the shark. Delap argues that even in the unlikely event that the author escapes mortal blows from such predators there are lesser monster critics to destroy the work.

Well into the eighteenth century, references were still predominantly to land sharks, although they began to take on more of the attributes of oceanic sharks. It is also a period in which we see more humour attached to the relationship between sharks and men. There are some direct references to oceanic sharks, for instance, in Captain Edward Thompson's masque, *The Syrens* (1776). In Act I the men have just experienced a storm and ensuing shipwreck during which the Commander is lost:

> Tom.
>
> I saw him sink, struggling to save his pretty, pretty Wife: – she was the daintiest thing that ever eat white biscuit: – her miseries might have charm'd a shark to have given her a tow to shore; but drowning, boys, is common to all. If there is no help for such heavenly workmanship – what chance hath poor merry fat Meg, my Lady's slipper –and yet, methinks, Meg would tilt light as an empty gin-cagg.
>
> Geo.
>
> No, no, she would leak fore and aft: Meg was scuttled, and not made for deep water; she will make a fine dish for a Dolphin.
>
> Tom.
>
> Special water *souchie* for a shark: well, make her merry; sighing is but windy diet, and we have lately had enough of that.
>
> (p.9)

The humour lies in the fact that, although it is their dear Commander that has been lost, it is his pretty wife that is the focus of their conversation. The perception that she could even have charmed a shark for a tow to land suggests both the extent of her appeal and the usual mercilessness of the shark. They conclude that if God cannot save this woman then He isn't going to bother trying to save fat Meg, a mere gin keg, since as George points out she is too leaky to float. In literature from the sixteenth century onwards, leaky women were perceived to be everything from promiscuous to talkative. Essentially, they were women who could not be contained (Shildrick, 1997). In both cases the women are symbiotically allied with the ocean in ways that men clearly are not.

Although there are several references in poetry and drama to dolphins not being friendly aides to humans, here there is a clear reference to the perception that they were man/woman-eaters. Tom's suggestion that Meg would make a 'Special water *souchie* for a shark' is also of interest. The *OED* does not reference 'souchie'. If what is intended is 'sushi', making this a humorous reversal in which the shark dines on raw flesh rather than the other way around, then it predates the *OED*'s first reference to 'sushi', in A.M. Bacon's *Japanese Interior* (1893), by a hundred years. We are reminded that before the arrival of mass media sailors were among the most knowledgeable about world cultures, as Captain Thompson would have been, from first-hand experience of them. Fat Meg does not, in fact, get eaten by either a dolphin or a shark but returns 'dripping like a Mermaid'. In his preface Thompson can be seen, as was by this time the fashion, comparing his play to a ship navigating a sea of dangerous critics: 'the Dramatic Cock-boat must take its chance on the Critic Sea!' (Thompson, 1776:5).

In Richard Cumberland's *The Impostors* (1789) there is a clear sense of the fusion of land- and oceanic sharks, when the two land sharks, Lord Janus and Polycarp, meet. Lord Janus upbraids Polycarp for being so shark-like as to eat shark, by which he means that Polycarp has turned on him, but he falsely represents this behaviour as that of cannibalistic oceanic sharks to convey this savagery. Again, in his play *The Jew of Mogadore* (1808), Nadab complains bitterly about having been stripped of all of his possessions by the Emperor, but he does so by fusing the idea of a land-shark and an oceanic shark.

> I wish you was further; tis' no great monies; only a few poor ducats in a poor canvass bag, that is all. But what then, were it a well, a well may be drawn dry; the sea itself hath a bottom to it. I am peeled, I am pillaged by these villainous Moors. I am eaten up alive by this shark of an Emperor, this Muley Ishmael, who regards the life of a Jew no more than he does that of a jackall.
>
> (p.10)

The Emperor doesn't merely take everything from him but eats him alive. Nadab conveys his desperate state by drawing on the common motif from poetry that we have already seen, that of being consigned to a living grave in the belly of a shark. In Act III Palti tells Rooney:

> I tell you, my dear friend, and you may take it on my word, the crocodile is not so false, the shark is not more savage, and the hippopotamos of the Nile is not less voracious for his prey than your old master Nadab for his pelf.
>
> (p.51)

As Charles Mabon has noted, characteristic of Cumberland's later work was his more favourable representation of minority groups. In both *The Jew*, an even greater success with audiences, and *The Jew of Mogadore* the Jews are generous philanthropists. Nadab is here misrepresented as a shark, as the shark in turn is misrepresented as savage.

Steven Mentz has argued that 'culture has always grappled with living in an unstable, ocean-drenched environment. Shifting our focus from the supposed stability of land, with its pastoral and georgic master narratives, to a broader vision that embraces the maritime world and what Melville calls "this terraqueous globe" will mean abandoning certain happy fictions and replacing them with less comforting narratives. Fewer gardens, and more shipwrecks' (Mentz, 2009:98). One of the more useful narratives he proposes we place at the centre of our studies is *Robinson Crusoe*. However, in Isaac Pocock's *Robinson Crusoe* (1817), the representation of the shark is not favourable. Pocock's is one of many dramatizations of Defoe's (1719) novel in the nineteenth century (Baines, 2007). Here, Glee and the chorus sing cheerfully about their piracy, drawing on the supposed behaviour of the shark to license it:

> When the anchor's a-peak,
> And the ship under weigh,
> The wide ocean we'll seek,
> Like a shark for its prey.
> We'll take what we can, boys,
> Wherever we steer;
> Friend or foe, 'tis all one
> To a bold Bucanier.
>
> (p.13)

This is an unusual conflation of the oceanic shark and land-shark at sea, making this crew the most dangerous aqua/terra creatures of all (Figure 4.1). In fact, far from being undiscerning feeders, different species of sharks have different diets, and even the same species have different diets at different stages of their development (Trystram et al., 2016).

Edward Jerningham's *The Siege of Berwick* (1793) tells the story of the clash between the English and the Scottish at Berwick in 1333. Sir Alexander Seton was the Deputy Governor of the town, left in charge at the time of the siege. He agreed terms with Edward III, as a consequence of which he had to hand over his sons as hostages. In this play, their mother, Ethelberta, determines to go to the enemy camp to beg for their lives. Warned against this rash plan by Juliana, Ethelberta insists:

> Talk not to me of dangers, I despise them.
> Say, hast thou not beheld the bold sea-eagle,
> When her dear young one from the rock hath fall'n,
> Descend undaunted to the roaring main,
> Dash with her throbbing breast the waves asunder,
> To snatch the nestling from the ravenous shark!
> And would'st thou have me pause and hesitate
> To act the mother's part?
>
> (p.44)

If a sea-eagle is brave enough to snatch back her chicks from the open jaws of a 'ravenous' shark then how could Ethelberta do less? This epic simile manifests an emotional struggle to convey what is otherwise invisible to the audience. It renders Ethelberta's internal struggle immediate and dramatic as she envisages her children in the jaws of death. It was understood at the time as a play about the political relationship between France and England in the wake of the executions of Louis XVI and Marie Antoinette (Darby, 1999). In Jerningham's simile the shark has not, however, actually attacked the chick at all, but the chick has fallen off a rock. The simile loses all of its power once we re-read it from the perspective of the shark. The shark's job of cleaning up floating, dead or dying debris is unhelpfully presented as horrifying. It not only represents the shark negatively, it undermines Jerningham's dramatic intention. He clearly means this to be a powerful contrast of maternal courage in seeking to rescue children that paternal cowardice has surrendered.

Thomas Holcroft, who is credited with bringing melodrama to the British stage (Garnai, 2010), also wrote *The Choloric Fathers* (1785) in the second act of which a song is sung that proclaims sharks as voracious, though positively restrained compared to land-sharks or lawyers:

> A Lawyer a nondescript monster we deem;
> Shark, Whale, or Leviathan's nothing to him:
> His green-bag's a belly which simile mocks,
> For it swallows up houses, fields, forests and flocks!
>
> (p.45)

As Stallbaumer notes, Holcroft's aim was 'to follow the English tradition of comedy and at the same time use the stage as a school of morality' (Stallbaumer, 1936: 35). Here, so monstrous is the lawyer that he cannot be represented, except as a consumer of all things. This lack of representation in a legal context also renders him indefensible. Of course, the shark is invoked here, albeit unfairly, as a measure of the lawyer's immeasurable greed.

John Baldwin Buckstone's *Jack Sheppard* (1840) is an adaptation from the novel about a notorious criminal by William Harrison Ainsworth. In Act II Scene 5 Jon assures Sir Rowland that he can soon get rid of his nephew by feeding him to the sharks.

> Jon *(Aside to Sir Rowland – handing him a pen, and placing the bond before him.)*
>
> What do you think of your nephew? Sign, and in a week his body shall feed the sharks of the North Sea!
>
> (p.33)

The advantage of feeding troublesome relatives to sharks is that they disappear without a trace, at least in dramatic fiction. Melodrama was an extremely important means of forging attitudes and 'played a significant role in shaping, articulating, and contesting changes in social and political relations, as well as in reshaping popular consciousness on the individual level' (Buckley, 2002:423). In reality, sharks have their own method of disposing of troublesome objects. They evert their stomachs through their mouths and then re-swallow them (Brunnschweiler, 2005). Humans are not the most easily digestible of prey for sharks, and human remains have been known to wash ashore, so feeding relatives to sharks may not, in fact, be the best way to get rid of them.

Although, as we have noted, references to oceanic sharks in the drama of the earlier period are sparse and on the whole stereotypical, in twentieth-century drama we meet with a much wider range of representations and they are often attached to a wide spectrum of human emotions.

Sharks and increasing emotional representations

Noel Coward's *Point Valaine* (1934) is a tense, sexual drama – a departure from the comedies for which he is better known. In Act 1 Scene 3 George blames Ted for being over-sensitive when he shows a fear of sharks. Ted's fear is what we could describe as typical, the fear of the unknown in the open sea. George reassures him that he is probably all right in the daytime, endorsing a popular myth about sharks only hunting at night. Ted invokes the story of a man who has apparently swum to the main island without being attacked, but what if he had been? Although George displays no sympathy with Ted's fear of sharks, he has to admit the reality of the fear of sharks and their potential threat to local economies. This is echoed in Act 2 Scene 1 when Linda and Martin, their friend, are swapping stories. Linda tells him that she had to choose the name of her hotel carefully to distract people from the fact that it was in a shark zone. In his introduction to the play, Sheridan Morley argues that '*Point Valaine* remains memorably the only Coward play in which a man literally throws himself to the sharks' (Morley, 1999:95). In fact, the broken-hearted Stefan leaps from a rock into the sea and drowns himself at the end of Act 3 Scene 1. Morley's memory is informed by an assumption that Stefan must have been eaten if there was a shark in the sea. In fact, we find out pretty quickly in Act 3 Scene 2 that his body has been found drowned and nibbled, probably by test bites, not eaten by sharks at all. If his intention really was to throw himself to the sharks, we can only conclude that they threw him back.

Season's Greetings (1980) is one of Alan Ayckbourn's most amusing and successful plays about human chaos. In Act 1 Scene 1 Ayckbourn creates the mayhem of Christmas preparations in inimitable style. Amid the confusion and cock-ups is Harvey, who is fixed in front of the TV, describing to everyone the killer shark fight that he is watching. Although there are over 500 species of shark, a killer shark is not one of them. Harvey's mistake is a sign of his excitement at watching two sharks fighting to the death, clearly not a documentary. What is interesting is the way in which the shark fight reflects something of the chaotic clash of human relationships around him as they are trying to pull a merry Christmas together. The juxtaposition of Belinda's focus on a Christmas tree decoration and Harvey's focus on sharks killing each other creates an incongruity of place, time and mood to humorous effect. It is important to note that although the assumption throughout is that adults are preparing Christmas for children there are no children seen in the play and no focus on Christmas as a religious festival. Each of the adults is self-absorbed (Tucker, 2006). This is a play about adults in their own, isolated worlds – and Harvey is at the centre of this remoteness in his voyeuristic fascination with two sharks attacking each other.

Human excitement over sharks is again apparent in *Seascape with sharks and dancer: a play* (1974) by Don Nigro, set in a beach house on Cape Cod. The play charts the difficult relationship between Ben, a librarian with aspirations to be a professional writer, and Tracy, a drifter, after he has rescued her, as he believes, from drowning. Tracy insists that she was in fact just dancing in the sea. After much solicitation from Ben, she finally tells her life story, more particularly why she has a deep connection to sharks. She explains that after a life of drifting, disappointment and final rejection from her beloved boyfriend she goes to the aquarium where she finally connects with something, and it is the sharks. However, sharks are not understood in a benign or loving way at all, but rather as ruthless killing machines.

Tracy displays several shark-like features she has observed, and even some she hasn't. When she refuses to acknowledge Ben's efforts to drag her out of the sea as a rescue, Ben offers her a thumb as a microphone so that she can report this abduction, which she then bites. Subsequently she tells Ben that she is pregnant but she imagines her unwanted baby as a fish rather than a child. Throughout the play, the depiction of sharks is as other-worldly, detached and cold, a pathetic fallacy, an externalisation, of Tracy's own sense of remoteness from human society. In one of his interviews, Don Nigro revealed: 'As a playwright, society often makes me feel as "marginal" and "useless" and "insane" as Columbus. If I couldn't identify with Columbus, I couldn't have written *Mariner*' (Constantinidis, 1993:149). In *Seascape*, Tracy is, of course, depicted as an extremely remote individual. We understand the analogy here because of other cultural representations of sharks as cold and alone in a vast, apparently directionless, ocean (Jacoby et al., 2012). However, the more we observe sharks in their own environments the more we are beginning to understand that they are not such quintessential loners.

Swimming with Sharks (2007) is an adaptation by Michael Leslie of George Huang's movie of the same name. It is a play that, among other things, deals with a complex love triangle. In Act 2 Scene 4 Dawn gets a phone call, thinking it is Guy, a promising young writer and her lover. It is Buddy, the movie mogul, and once he realises that he is not going to be able to seduce Dawn he lets her know that Guy has merely been using her for his career, pretending to be a seal-cub while he is in fact a shark. If Guy is the seal cub, Buddy is surely the shark in this scene as he attacks Guy to gain Dawn's focus. What is interesting in this appropriation is the way that the lethal scenario of a shark and seal, the favourite food of some of the larger sharks in an oceanic environment, is used as an epic simile to create emotional energy within the scene, which it would otherwise lack as an onstage telephone call.

Sharks are also used amusingly in twentieth-century drama, for instance in April de Angelis' *The Positive Hour* (1997). In Act 1 Scene 7 Paula and her eight-year-old daughter Victoria are chatting with Miranda, their case officer. Having abducted Victoria earlier in the day from her foster home, Paula now relates their day's adventures – the highlights being *Baywatch: The Movie* and Burger King. Victoria asks why the drowning man in *Baywatch* wasn't eaten by a shark. In an amusingly desperate effort to answer the question, Paula suggests that sharks are too expensive. Victoria then begins to eat the left-over chips whole and with great gusto, evoking the absent shark.

The scene provocatively portrays a young child's perception of shark behaviour from other cultural representations. Clearly Victoria's question about the absent man-eating shark reveals disappointed expectation based on what she has been told about them so far. Paula's answer incorporates them in an economy relationship with the movie business that appears absurd, but it also points to a certain perception about their persistent exoticism. The fiction has to be believable to the child. The question of the absent shark amusingly reflects the awkward negotiations between parent and child as they both strive to make sense of the world.

In another more recent play, an autistic savant child offers a perception of a shark that is strangely in accordance with what has by now been established as the perceptual norm. Simon Stephens's *The Curious Incident of the Dog in the Night-Time* (2012) is based on Mark Haddon's novel (2003). In Part 1, Christopher, a boy of 15, relates his experience of being in the sea. At first he says he doesn't like swimming because he doesn't like taking his clothes off, but this excuse is quickly replaced by what is evidently a more ingrained fear of sharks. As he watches his mother disappear under the water, he immediately assumes that she has been eaten by a shark. He begins to scream. She resurfaces and assures him that Cornwall doesn't have sharks. This is an interesting 'episode' in a play about a young boy's alternative experience of the world. What first prompts his episode is the suddenness of his mother's disappearance. He doesn't assume she has drowned, although there is a much higher chance of this happening. In his mind the sea engulfs his mother in exactly the same way that a shark would. Her reassurance fails to take notice of his response to a sudden absence. Ironically, it is a false assurance. Many shark species spend time in the Atlantic, including the shark that was ushered into the English language in 1569 (see Chapters 1 and 2). The Cornish even have had their own word for it (see Chapter 2). To date, only one 'shark attack' in the British Isles has been reported, on 7 June 2017, off the Cornish coast, which resulted in minor scrapes and bruises, but which predictably received wide media coverage.

Fear and reason are again pitted against each other in *Shoreditch Madonna* (2005) by Rebecca Lenkiewicz. There are several references to sharks throughout the play. In Act 1 Scene 2 Michael, Nick and Hodge are setting up cameras for a shoot. Eventually, the footsteps of a has-been artist, Devlin, are heard as Michael alerts us to the ambiguity of the man. Is he a shark or a pussycat? Both descriptions suggest 'slipperiness' of a kind, but the implication is that one is lethal whereas the other is benign. The small room in which they are shooting then becomes a focus of attention. Devlin demonstrates that he has overheard Michael by measuring the smallness of the room in sharks and cats. A conversation then ensues between Devlin and Hodge about how terrifying and restless sharks are, invoking the long use of sharks in literature as swimming coffins. These old misconceptions about sharks are scientifically validated by Michael, though they have little basis in science. The exchange usefully demonstrates some of the processes by which cultural representation comes to represent fiction, in the absence of fact, as truth.

Lenkiwicz said of *Shoreditch Madonna* that it was 'a strange creature that came out of the dark . . . A diehard artist negotiates life in the shadows after being expelled from the bright lights. Who is in? Who is out? Who merely has to pixelate to gain plaudits? Real versus slick. It is about love and the desire to connect' (Lenkiewicz, 2013:8). The difficulty of human relationships and the remoteness of people from each other are signalled throughout the play by the idea of the lone shark. Devlin has a real and repeatedly referred to fear of sharks. In Act 2 Scene 14 the reference to the oceanic sharks becomes more specific. Martha and Devlin are seen on a street with a young boy, Tom, a consequence of their one-night stand, who then runs off. After a brief discussion, during which Martha explains how important Devlin is to Tom, Devlin agrees to spend more time with him. However, his fear of sharks is again made explicit when he refuses to take him to the aquarium. Devlin describes his fear of sharks, but his description is of a large great white shark. Martha tries to reassure him that they are contained, to no avail. Most interesting about the tracing of Devlin's fear of sharks is the way that it ignores reason and logic. The sharks are contained, and are highly unlikely to be great white sharks. Large pelagic sharks are very difficult to keep in aquaria and have a bad track record of survival in tanks. Smaller species or baby sharks are more generally acquired as exhibits. Devlin's imagined shark is, however, the shark of a cultural construct – the shark of *Jaws*. No attempts at persuasion will convince Devlin that he has no reason to fear sharks, but it is this fear that in the play keeps Devlin in check and makes him human, after all.

Rebecca Lenkiewicz's later play *The Painter* (2011) is about a more successful artist. In Act 1 Scene 5 the painter William Turner tells his model Jenny that he has a good feeling about the year. Jenny tells him that there is always discussion about abolition and reform, but it never happens, and it is not possible to fish all of the dead out of the sea. This is a preoccupation in the poetry and prose of the eighteenth and nineteenth centuries, already observed. Here it stands in for all the violence done and the impossibility of undoing it, even with new legislation. 'Shark-infested' has become a popular means of referring to any oceans with sharks in them, but, as divers repeatedly warn us, numbers are becoming so depleted that the term 'shark-infested' is becoming obsolete, or extinct, itself. Of course, Jenny's point here is that because the captains ensured that the slaves would drown, and, she suggests, probably be eaten, there has been no historical accountability. The absent bodies are absent evidence that ensures that they will never receive justice. The play further testifies to the relationship between sailors and sharks as they travelled the oceans together, even though, as we know, sharks that followed ships were often not regarded as friends.

Sexy sharks

One of the more unusual representations of sharks in drama is as a focus for sex, an example of which we find in *Woman and Scarecrow* (2006) by Marina Carr. A woman who is nearing death reflects on the moments of her life that only now appear to fall into meaningful context. When she is asked by Scarecrow about one of her lovers she suddenly realises that she had unknowingly become a substitute for a shark in a sexual fantasy. In Act 1 the Woman remembers not merely that he watched breaching shark footage before he had sex, but that he replayed it and froze frames of its jaws. The shark appears to symbolically encompass both male and female sexual power. The image of the vagina dentata or vagina with teeth finds representation in ancient cultures across the globe, with different semantic associations. While the woman clearly understands this as castration anxiety, the cowboy associates it with phallic energy that triumphs over the symbolic other in his sexual consummation. However, the only eruption dramatized is that of realisation on memory. It is a clever use of sharks to convey our complex emotions surrounding them, from manifest fear and sexual anxiety to the ultimate expression of power.

In Jez Butterworth's *Parlour Song* (2008), a suburban one-act play about a demolition expert, his wife and their next-door neighbour, who is a car-wash manager, several mundane games are introduced through which we see relationships weave together, with the shark again appropriated for sexual purposes.

In one scene, Joy and Dale, her lover and neighbour, playfully exchange stories. Joy asks Dale whether he would rather be eaten by a shark or a lion. Dale chooses the lion on the grounds that the lion kills you outright and then shares what he eats with his family, whereas the shark repeatedly bites you to death. Joy chooses the shark because she wants to be torn apart as she watches her blood seep into the water. Dale's perception that death by sharks is like being rushed at repeatedly is already a warning to Joy to cease. Dale prefers a gentle mauling in a family setting. Joy appears to have picked the wrong target. The perception of the shark violently pounding and tearing is sexualised as a most complete and final abandonment. Joy does not fantasise about the much more probable bleeding to death from a single test bite. Her shark fantasy is also fundamentally about her desire to destroy the stale domesticity that binds her. As this scene promises, Joy's plan to run off with her neighbour's husband fails miserably, and she returns to her former life – a failed heroine.

A tenacious fascination with oceanic sharks is again apparent in Rona Munro's *Long Time Dead* (2006). In Act 1, Gnome, Grizzly and Dog, three climbers, are caught on the side of an ice cliff. Gnome is injured and Grizzly and Dog come to her rescue. In order to deflect from the horror of the situation, they engage her in a series of conversations, beginning with Keith Richards and Rock and Roll. Soon, their conversation veers towards the worst ways of dying. Gnome decides that being eaten alive by a polar bear would be worst. Grizzly suggests that they are the most dangerous animals of all, which, being called Grizzly, he feels empowered to say. However, Dog suggests that we must be the most dangerous animals, since polar bears are actually endangered. All the while, these two are slowly manoeuvring ropes in an attempt to rescue Gnome. Dog then suggests that getting caught in a tidal wave would be worst. Grizzly tells them that his friend, Hector, was eaten by a shark. It quickly emerges that he was not eaten, but that one of his legs was bitten. What ensues is a description of how Hector felt as he was being manoeuvred from beneath the waves. Contrary to most movie representations of shark attacks, screaming in agony is not the usual response. As Grizzly notes, it is the sudden awareness of an involuntary movement of the body that alerts the unfortunate person to the fact that something is wrong. This would not make emotionally impactful movie scenes so it is an encounter fact that is generally completely disregarded.

As Dog and Grizzly move onto the next worst way to die, being caught in a volcanic explosion, a fascinated Gnome asks Grizzly to finish the shark story. Throughout this desperate scene, during which they try to rescue a badly injured climber, they consider worse situations to be in as a means of keeping up morale and taking their minds off the horror and stress of the situation, but the shark scenario is the only one that has a story attached to it and that is intriguing enough to be returned to after all of the other worst ways of dying have been considered in turn.

The gruesome fascination with the fantasy of the shark eating Hector is crucially juxtaposed with Hector's alternative real experience. The story hooks the imagination of the listeners, even while one of them is dying, and is sapped of emotional drama with the simple relation of Hector's own account. Speculation ensues about how Hector managed to escape the shark, for Hector said nothing. It is a useful scene that exposes our propensity to exaggerate events and the ease with which we supply any gaps in information with fictions that are then perpetuated.

Human conflict is at the centre of human relationships in Philip Ridley's *Leaves of Glass* (2007), in which two brothers continually renegotiate their relationship in the context of abuse. As Andrew Willey points out, it is a play that focuses on 'interplay of past predatory seduction, of psychological damage and of covering up the truth' (Willey, 2013:73). Scene 5 opens with Barry apologising to his brother Steven for letting him down. He blames what is often referred to as 'sharky' weather. This becomes an underlying theme throughout the scene as Barry explains that the reason he couldn't carry out the graffiti removal job was because he was so inspired by it, even though it was of a bomb exploding in technicolour. Barry explains how he ran for his life from the crowd that was admiring this gruesome artwork.

Barry describes Steven as restless through the extended analogy of a shark, and then humorously adds that it may or may not be true that sharks are always in motion, but it is what he wants to believe. The scene ends with Barry finally quitting, leaving Steven to confront his own guilt.

It is an interesting scene in many ways. Barry is an artistic character who responds to life, and colour and energy, with excitement. Steven is his nemesis, a businessman whose company had developed a water-blasting device that erases graffiti, colour, energy and memory from street life, and who as teenager is responsible for having prostituted Barry. Sharks supply levels of meaning throughout the play, informing character, mood and plot. Clearly, the idea of a predatory shark is a powerful indicator of this fraternal history. The false apprehension that sharks never sleep is perpetuated, as we have seen, throughout poetry and drama to signal the otherness of sharks; here it is specifically their energy that appeals to Barry. The scientific truth of the matter is we do not know whether sharks sleep in the way that other animals sleep, although we suspect that they may show circadian changes in activity. However, Barry draws our attention to the fact that people believe what they want to believe. Fiction is often much more exciting, and easier to live with. Barry doesn't want to live in a world without these fictions. The shark represents this restless energy that makes him want to live – and makes him want to quit the business of erasing colour from plain walls.

In Scene 15, Steven signals his diametrical opposition to Barry's character when he reveals that, for him, the shark represents not energy and life but anonymity in the darkness, privacy and the retreat that he craves. It is a useful play in highlighting the ambiguity of the shark and the incessant need that humans have to adapt sharks for their own purposes. In this play there could not be two more diametrically opposite character types than Barry and Steven, for whom sharks have a very different allure.

Childhood and sharks are again encountered in Sam Holcroft's *Cockroach* (2008), a play about growing up in a global war context. It opens with a secondary school Biology class in process and a teacher, Beth, trying to control a hysterical student, Lee. Amid the chaos Beth attempts to teach evolution. Beth begins her lecture about the endangered listing for basking sharks and cheetahs but is interrupted by Lee who has noticed trucks on the playing field – which he is much more interested in. This highlights an important problem in the representation of evolution, conservation and ecology in the classroom.

When the play was first staged at the Traverse Theatre, Edinburgh, the critic Mark Fisher suggested that within the play was a suggestion that 'war provides a necessary cull of male aggression, paving the way for some non-violent future' (Fisher, 2008, *Guardian* online). However, it is not necessary to read the play as a lesson in destruction by testosterone. The chances of any of the children in this classroom ever coming meaningfully close to a cheetah or a basking shark are very small indeed. The basking shark stands in for that which is inaccessible and remote from their childhood experience. The task is to make the truth about sharks meaningful through wide-ranging strategies that appeal across the spectrum of human society, and this is a cross-disciplinary venture.

Robin French's *Gilbert is Dead* (2009) had its debut at Hoxton Hall in London. It was snubbed by Michael Billington, *The Guardian*'s long-time theatre critic, as an eccentric treatment of the Darwinian debate (Billington, 2009), but it has rather more than that to offer. In Scene 2 we gather that Lucille's father is a brilliant taxidermist. Meriwhether is in the process of trying to take over the museum and remove the sixteen-year-old Lucille from its confines. However, there is little that Lucille doesn't know about life from the contents of this museum, and in the process of demonstrating this she relates a story of how all sea-life came to be monstrous. Her description is of how sea creatures, sharks, sea lions and whales came to be the culturally constructed monsters of horror stories.

What is important is that Lucille is able to demonstrate that she can understand the difference between imagined horror and real horror, and that real horror is not to be found in nature, not even in her subsequent lurid description of the Marabou Stork, but in the worlds that men imagine, create and build for themselves. As we have so often seen in the poetry and drama, monstrous sharks are actually imagined externalised manifestations of monstrous humanity.

Jo Clifford's *Every One* (2010) is an imaginative adaptation of *Everyman*. Clifford's plays challenge the idea of normative existence. Maddy Costa suggests that '*Every One* is disgusted and incensed at the damage inflicted by capitalism. This is a play in which death claims responsibility for all the ills of the world: war, genocide and – his 'masterstroke' – an economic system that kills those it benefits as much as those it crushes, but he does so in the body of a man' (Costa, 2016: Preface). More important for us here is that it demonstrates the contradictions inherent in human relationships with sharks. In Act 1 the family is gathered in the living room. Mazz, the daughter of the family, shares a dream she has had about fighting off sharks. Having beaten them back with a pepper pot, she has to take cover when Kevin comes into the room with his virtual reality game – and begins to hit first the sharks and then Mazz. She protests that she is not a shark and then sees that she is. Joe offers an interpretation of the shark dream, informing Mazz that the local fish counter is currently selling shark steaks. Indeed, that Mazz should first attack the shark with a cruet is one possible explanation of the dream connection. However, Joe has a more pressing argument to make that once animals have run out, humans will be on the menu. The point is well made, and it is very rare in drama to find that the killing of sharks is the focus of conservation outrage.

This contradiction in the system, in which conservation is being led by consumption, has been well documented in ecological studies in the last decade. The anticipated ecological outcome of overfishing, although not as extreme and absurd as that advanced by Joe here, is collapse. Joe wonderfully imitates the packaging and marketing of live animals as breakfasts, lunches, dinners and snacks, and gets our attention when he envisages a time when humans themselves will be harvested. We are reminded that Mazz's heroic struggle with a shark is but a dream; the reality is that the sharks are now steaks in a fridge.

In a more humorous reference to shark culture, in *Nativities* (2012), Zoe Cooper explores belonging and isolation in the artificially co-operative context of an office. In Scene 12 Madge is seen knitting while Stella looks down at the sandwiches in her Tupperware box. Stella tries to convey to Madge that she is not happy about the fact that her pregnancy has been circulated to the whole office and that she definitely doesn't want an office party to celebrate it. Madge senses her unease and tries to explain that it isn't a big thing – but that they try to celebrate for everyone in the office. The way she explains this is by reference to her son's birthday party. She lovingly describes the detail of the design and decoration of the shark cake that she made for her son as a means of conveying to Stella an obvious contrast to the kind of celebration Stella can expect.

What is interesting about Madge's shark, covered in smarties, is that it bears very little resemblance to an oceanic shark. It is a loving mother's edible representation of a shark, which in an iconoclastic moment turns 'Jaws' into a colourful, toothless, chocolate shark with an anatomical addition of a liquorice tongue as an extra treat. Madge explains that the cake is a monument to her son's favourite time with his dad, which is when they go shark fishing. Stella's party will have no such yummy shark. It is intended merely as a gesture of good will.

Finally, in our selection of recent plays, David Hare's *The Moderate Soprano* (2015) has the shark in a suit, a land-shark, but throughout he is represented by the oceanic shark. The play is organised by periods, though not always in sequence. In Scene 4, set in 1934, to a considerable degree Rudi literally looks like a great white shark in a suit. In Scene 20, set in 1962, John addresses the fact that Rudi is a loner, again relying on the myth that sharks are non-social. Rudi's sometimes necessary ruthlessness in sacking his singers is enabled only by his refusal of their friendship and society, suggesting that this is merely his strategic defence against the pain of broken friendships. However, his silence makes him a dangerous and unknowable entity. As Michael Mangan observes, Hare's plays often deal with power struggles in well-established political contexts with a clear focus on an 'alpha male': 'the alpha male in Hare's plays is often the figure who most clearly represents the relevant power elite and its values, whether that power is seen in financial, political or cultural terms; he is also characterised, as is the typical alpha male in real life, by a high degree of self-confidence' (Mangan, 2007:222). The oceanic shark is keyed into the world of opera, based on the real story of the founder of Glyndebourne Opera, John Christie. Rudi is based on Rudolf Bing, the manager of the Metropolitan Opera (New York) for most of his earlier career. He became General Manager of Glyndebourne between 1936 and 1949. The comment 'the smoothest shark in the water' reflects the awe that he inspired, as well as testifying to his mastery of a challenging environment – deserving of everyone's respect.

References

Primary works are derived from Chadwyck-Healey Literature Collections online unless otherwise stated.

Baines, P. (2007) *Daniel Defoe: Robinson Crusoe/Moll Flanders.* London: Palgrave Macmillan.

Billington, M. (2009) Gilbert Is Dead: Review. *The Guardian online.*

Brunnschweiler, J.M., Andrews, P.L.R., Southall, E.J., Pickering, M. and Sims, D.W. (2005) Rapid Voluntary Stomach Eversion in a Free-Living Shark. *Journal of Marine Biology*, 85, 1141–1144.

Buckley, M (2002) 'Jack Sheppard' and the Mass Audience. *Victorian Studies,* 44.3, 423–463.

Constantinidis, S.E. (1993) Playwriting and the Masks of History: An Interview with Playwright Don Nigro. *Journal of Dramatic Theory and Criticism*, 8.1, 149–156.

Costa, M. (2016) *Every One (*Preface) in Clifford, J. 2016. *Every One.* London: Chris Goode & Company and Battersea Arts Centre.

Cressy, D. (1997) *Birth, Marriage and Death.* Oxford: Oxford University Press.

Darby, B. (1999) Spectacle and Revolution in 1790s Tragedy. *Studies in English Literature, 1500–1900*, 39.3, 575–596.

Freeman, T.M. (1995) *Dramatic Representations of British Soldiers and Sailors on the London Stage, 1660–1800.* Lewiston, New York: Edwin Mellen Press.

Garnai, A. (2010) 'A lock Upon My Lips': The Melodrama of Silencing and Censorship in Thomas Holcroft's 'Knave, or Not'? *Eighteenth-Century Studies*, 43.4, 473–484.

Jacoby, D.M.P., Croft, D.P., and Sims, D.W. (2012) Social Behaviour in Sharks and Rays: Analysis, Patterns and Implications for Conservation. *Fish and Fisheries,* 13, 399–417.

Lenkiewicz, R. (2013) *Plays 1.* London: Faber.

Mabon, C.B. (1899) The Jew in English Poetry and Drama. *The Jewish Quarterly Review*, 11.3, 411–430.

Mangan, M. (2007) 'Marbled with Doubt': Satire, Reality and the Alpha Male in the Plays of David Hare. In Boon, R. (ed.) *The Cambridge Companion to David Hare,* Cambridge: Cambridge University Press. 220–235.

Marten, R.A., Hammerschlag, N., Collier, R.S. and Chris Fallows (2005) Predatory Behaviour of White Sharks (*Carcharadon carcharias*) at Seal Island, South Africa. *Journal of Marine Biology Association, U.K.* 85, 1121–1135.

Mentz, S. (2009) *At the Bottom of Shakespeare's Ocean.* London: Bloomsbury.

Milne, A. (2015) The Place of the Poet in Place: Reading Local Culture in the Work of Mary Leapor. *Tulsa Studies in Women's Literature.* 24.1, 125–139.

Morley, S. (1999) *Coward: Noel Coward Collected Plays: Six.* London: Methuen.

Shildrick, M. (1997) *Leaky Bodies and Boundaries.* London and New York: Routledge.

Stallbaumer, V.R. (1936) Thomas Holcroft: A Satirist in the Stream of Sentimentalism. *ELH*, 3.1, 31–62.

Tucker, S. (2006) Diptych of Comedy and Carnival: Alan Ayckbourn's *House & Garden. New Theatre Quarterly*, 22.2, 155–180.

Thompson, E. (1776) *Syrens, A Masque, in Two Acts.* London: Kearsly.

Trystram, C., Rogers, K.M., Soria, M. and Jaquemet, S. (2016) Feeding Patterns of Two Sympatric Shark Predators in Coastal Ecosystems of an Oceanic Island. *Canadian Journal of Fisheries and Aquatic Sciences,* 74.2, 216–227.

Wyllie, A. (2013) Philip Wyllie and Memory. *Studies in Theatre and Performance*, 33.1, 65–75.

5 Sharks in the visual arts

Depictions of sharks have differed so sharply between Eastern and Western art that we may be forgiven for doubting whether the twain, to use Rudyard Kipling's words, would ever meet. The Western tradition, comprised of artists on both sides of the Atlantic, is overwhelmingly sensational, exaggerating the horror, the size, and the toothy jaws of sharks, making them monsters from our nightmares rather than rendering them in realistic terms. In Asian and Pacific art, however, sharks have been presented in more accurate and commonplace terms, without spectacle. The reasons for this divergence in traditions are several, deriving first from familiarity, or lack thereof. For Pacific islanders, sharks have always been part of the given, more plentiful than wolves were for rural Europeans or bears for American settlers. Polynesians swam with sharks and frequently counted them among their gods, and even believed them to be inhabited by the spirits of their ancestors. Shark teeth and sharkskin were incorporated into everyday artefacts. But for both Europeans and North Americans, sharks remained rarely seen and poorly understood, and inaccurately depicted until late in the eighteenth century, despite early attempts to represent shark attacks dating back to an illustration on a vase from 725 BC (http://www.sharkattackfile.net/). The slave trade, specifically the hyperbolic lore surrounding it, and the chronicles of Captain Cook's voyages to the Pacific, were to have a lasting effect on the representations of sharks in the visual arts.

Shark depictions in Western art derive not only from centuries of shark slander, fully documented in an earlier chapter, but also from a long-held belief in an underworld and in demons. Whereas Eastern and Oceanic belief systems lack any conception of hell or the Angel of Darkness or eternal damnation, Christian religions regard these embodiments of absolute evil as living truths, thanks not only to the Bible but even more vividly to Dante, Milton, and two millennia of imaginative theologians. The Western imagination is also deeply invested in the myths of sea monsters, whether dragons from the deep or prehistoric megalodons or Jules Verne's giant squid. The major encyclopaedia of the Renaissance, Conrad Gessner's *Historia Animalium* (1558), featured beautifully accurate representations of plants and land animals, but presented sharks as eel-like things with upturned snouts, looking more akin to the creatures from the lightless depths, benthic rather than the pelagic sharks they would have encountered (Figure 5.1).

One reason for this distorted representation may be that these shark illustrations were drawn from dried or desiccated carcasses (Ellis, 2012:58). But mythical descriptions of sharks, or deliberately symbolic stylisations, also eclipsed realistic depictions. Maritime sailors did not, as a rule, paint, and major artists did not tend to shark-fish. Until the late nineteenth century, very few artists of renown or skill had even seen a shark. However, the exceptions are notable, both for their realism and the relative calm of their shark depictions. John White, one of the founders of what came to be known as the Lost Colony on Roanoke Island, North Carolina, painted the documentary portrait *The Manner of Their Fishing* (1585), depicting a group of Native Americans using spears and traps to catch fish, including hammerhead sharks (Figure 5.2). White was attached to the expeditionary colony as a 'recording artist' an amateur artist and mapmaker charged with 'drawing and mapping the peoples of land' (https://encyclopediavirginia.org). He is better known as the grandfather of Virginia Dare, the first child born in the New World. The British naval officer and amateur artist George Raper drew a beautiful if undetailed watercolour entitled *Port Jackson Shark* during a voyage to Australia in 1789 (https://fineartamerica.com), and accompanying Captain Cook on his South Seas voyages in the 1760s and 1770s were several illustrators, including the talented if little known Sydney Parkinson, who produced accurate watercolours of the sharks that they encountered, in particular a very beautiful blue shark that he rendered in full detail. The shark that Parkinson entitles *Squalus carcharius* is so accurately drawn that we recognize it as an oceanic white tip, *Carcharhinus longimanus*, with its distinctive rounded dorsal fin topped by a big blotch of white. Parkinson would also have recognized this dangerous shark, but there was not yet a lexicon for identifying it as a separate species from the dozens of other carcharhinids or requiem sharks, hence the watercolour's vague title.

In the early 1800s, various anonymous artists, likely amateurs on merchant ships, created wood block prints that also portrayed sharks caught at sea: undetailed but roughly accurate sharks, including a large one dangling dead off the high stern of a ship. It is worth noting that apart from the size of these documentary sharks, there's nothing alarming about them, no ferocious poses or emphasis on teeth. Furthermore, since these sharks are typically dead, they seem all the more tranquil and un-dynamic. The artists aboard ship with Captain Cook and on merchant ships were amateurs and part-time practitioners. Their aim was merely to record, not elaborate. These artists were tasked with reproducing the strange flora and fauna of new lands; they were not celebrated members of art academies able to sell their work to collectors. Thus, in terms of shark portraiture, they were the exceptions, not the norm.

The professional norm included not only the fanciful imaginers of sharks as sea monsters, but also the pivotal artist John Singleton Copley. His painting *Watson and the Shark* (1778) is an acknowledged masterpiece, a bridge between the great classicist painters like Michelangelo and the realists (Figure 5.3). It is also a thrilling portrait in its own right. Copley was both a sensationalist and a dramatist, and much more than either term implies. His work was more intricately historical and attentive to the moody features of common men than his audience and wealthy patrons were likely to understand. How Copley came to create this great work is in itself a great story, drawing together British colonialism, the American Revolution, and the natural history discoveries made by Captain Cook, who was for a brief but crucial period Copley's neighbour on London's Leicester Square.

Watson and the Shark

John Singleton Copley was born and grew up in Boston, Massachusetts, one of the major ports for commodities shipped in the great clockwise circle created by currents and prevailing winds around the Atlantic. Tea and finished goods came from London; slaves from West Africa; sugar, molasses, and rum from the Caribbean; tobacco, timber, turpentine, and later cotton from the American colonies. Copley's family was Irish and their circumstances modest, but when his mother remarried, their social station sharply improved. Copley showed himself to be a talented portraitist, and he married an heiress, Susanna Farnham Clarke, from Beacon Hill. His wife's family was not just rich but loyalist; thus, at the time of the American Revolution, they decamped from rebellious Boston to London's upscale Leicester Square. For young Copley, as well as for realism in art, that location was fortuitous.

Watson and the Shark debuted in 1778. In the Royal Academy's exhibition catalogue for that year, Copley's painting bears the more cumbersome title: 'A boy attacked by a shark, and rescued by some seamen on a boat; founded on a fact which happened in the Harbour of the Havannah' (Neff and Weber, 2013: 163). In 1749, while serving as a seaman's apprentice, Brook Watson was swimming in Havana harbour when he was attacked by a shark three times. Sailors in a rowboat eventually rescued Watson, who lost his lower leg but recovered and went on to a very prosperous career as a trader, working with such important British merchants and military figures as Joshua Winslow and General James Wolfe. He became the famous Sir Brook Watson, with his equally famous wooden leg, a fact confirmed by Watson himself in a newspaper account of the time. In later years, Watson was quite voluble in his memories of the experience, describing 'the afflicting sight of him seized and precipitated down the flood with his voracious assailant' and his salvation 'despite all the flesh was stripped off the calf from the bone downwards and a foot . . . divided from the leg' (Neff and Weber, 2013:164–5). Thanks in part to the publicity, the painting was a huge success, and won Copley an election to the Royal Academy (Neff and Weber, 2013:164).

Copley's painting harkens back in idealised anatomical detail to Da Vinci's depiction of the young male nude. It also anticipates the great seascapes of J.M.W. Turner with its ships' masts and stormy sky. Even more importantly, the painting helped to inaugurate a new trend toward realism by depicting in detail the ordinary seamen struggling to rescue their shipmate from the shark's toothy jaws. Copley depicts a thrilling scene, a still moment captured in the midst of movement. The seamen's faces are individually detailed and their rough, homespun clothing rendered with its pleats and folds. Although stylised, the shark in Copley's painting displays a fully convincing pectoral fin and scythe-like tail, along with the olfactory openings on the shark's snout, and one small eye with the vertical iris characteristic of the bull shark – the likely culprit in any Caribbean harbour. The shark's open jaws provide a static moment in a kinetic scene, showing the bite about to occur, and as we look closely we note the evidence of at least one previous bite. Blood spirals away from the boy's lower left leg, and there's no foot to be seen. What we do not note, unless we're informed, is the resemblance between these shark jaws and those featured in museums and other displays that Copley had taken an interest in. Before Copley, European art too often relied on sea monster myths as representations of sharks, with arms and legs like salamanders and the goat heads of satanic demons. Even Gessner's authoritative encyclopaedia identifies sharks as 'sea-dogs' and 'dogfish' and depicts them as various dragon-like sea monsters. Of the hammerhead or 'sledgedog', Gessner fancifully writes:

> This fish shall be a very large, beastly and terrible animal, which never goes to shore because there it could catch only small prey. It eats all kinds of fish, and will also swallow and tear apart swimming people. When sighted, it is considered a sign of hateful bad luck.
>
> (Conrad Gessner, *Historia Animalium*, 1558)

What informed Copley's art was, first, his humble background among sailors and slaves on Boston's Long Wharf and, second, the natural history museums that he had been privileged to visit. On his grand tour Copley had actually seen a bull shark's jaws at the museum in Heidelberg. He was particularly influenced by the newly created natural history museums of Heidelberg and London and the smaller but no less important personal collection of Copley's neighbour on Leicester Square, Captain James Cook, who was back in London between his voyages to Polynesia. In Leicester Square he had viewed the documentary paintings and sketches by Cook's illustrators, including Sydney Parkinson. Copley could also inspect actual artefacts, masks and weapons and ceremonial costumes that Captain Cook returned with as appropriations or souvenirs. On his last, fatal voyage to the Pacific, Captain Cook even attempted to bring back with him a living artefact, a Hawaiian king, as did earlier and later English explorers with Princess Pocahontas from Virginia and the family members of African chieftains. Leicester Square contained no such living relics besides Cook himself, but it was still a historic intersection of exploration and art, enabled by coincidental proximity.

Copley crossed social class as well as an ocean to arrive in Leicester Square. Boston's Long Wharf had provided him ample opportunity to observe the sailors in detail, and growing up as a non-patrician, he was more inclined to do so. Looking again at *Watson and The Shark* we see that the black sailor is central in Copley's scene. We note, too, that every other figure is in motion: the sailor (perhaps a young officer) is poised to stab the shark with a pike, the shark is in liquid motion, the stricken Watson reaches up to be saved, the two sailors stretch out to save him, and the old sailor drags on the shirt of one of those sailors to keep him in the boat. Even the background sailors appear actively huddling in sympathy and consternation. Only the black sailor stands alone. He registers shock at the spectacle playing out before him, but he doesn't actively participate in the rescue. One critic at the time referred to him as an 'idle Black,' paralyzed by dread of some primitive superstition (Kamensky, 2016:282), a characterisation that we recognise as both condescending and presumptuous. Standing at the apex of the arranged figures in the boat, the black sailor is in no position to reach over his shipmates and grab either the stricken boy or a harpoon. Instead he stands witness.

In Copley's day, stories of sharks following slave ships across the Atlantic were so much repeated that they became lore. Europeans and Americans who had never seen a shark had heard the tales of the dead and dying slaves being tossed overboard to the ever-present sharks, of rebellious slaves also being thrown to the sharks, and of escaped slaves devoured by sharks as they tried to swim ashore in Caribbean ports, including Havana Harbour. The drama of the shark's attack on Watson is thus informed by the presence of the black sailor and magnified by it. Copley reminds us of the Middle Passage, of the slave ships crossing from Africa to the Caribbean and the American South, and of a history that, once we recognize it, haunts the painting.

Viewers in 1778 would likely shiver even more than we do because of their deeper enthralment with the sublime, the pleasure and beauty inherent in terror. In Edmund Burke's 1757 treatise *On the Sublime and Beautiful*, he asserts that 'terror is in all cases whatsoever, either more openly or latently, the ruling principle of the sublime' (Raban, 1993:9). Earlier eighteenth-century essayists also theorized about the thrills of chills, with John Dennis writing in 1704 about 'enthusiastic terror' (Cohen, 2010:113) and Joseph Addison asserting the existence of an 'agreeable horror'. Indeed, the eighteenth century gave us great gothic romances, but visual art was slower to follow.

Returning to the black sailor in *Watson and the Shark*, we note that his face is beautifully detailed, not generic as one or two of the other sailors are. Copley painted three versions of Watson and the Shark, one hanging in Washington, the more famous horizontal version, one in Boston and one in Detroit. The reproduction in this volume is the vertical version from the Detroit Art Institute, revealing Watson's leg bleeding copiously, and missing a foot, in the murky water. The vertical orientation of the painting invites our eye to travel the depths of the ocean and its dangers. The shark, seen returning on the surface, also signifies the mythical monster that we fear might emerge out of the gloom. Whether Copley intended to evoke the sublime in his second version is a question that we can't answer. Copley may simply have been inspired by the vivid possibilities of realism – or the kinetic and commercial potential of his subject. On the other hand, Edmund Burke had published his thesis on the Sublime in 1757, meaning that his revolutionary ideas were in the air.

The sublimation of the sea

Early nineteenth-century illustrations and engravings leaned heavily on the popular motif of large sharks and small boats. M. Dubourg's *Killing a Shark* (1817) is an obvious imitation of Copley's painting, showing a crew of sailors in the act of harpooning a menacing shark about to attack a young woman wearing what appears to be a nightgown. *Sharks in the Adriatic*, an anonymous engraving from *Le Petit Journal* (1908), shows three girls in full Victorian garb on a boat trying to fend off sharks with oars while a fourth girl is half in the boat and half in a shark's toothy maw. An untitled engraving by Castelli (1878) depicts a fisherman, also in a small boat, pulling a child aboard in the nick of time to save her from the jaws of a shark with a mouth shaped like that of a large-mouthed bass. Other, more accurate, documentary prints show dead sharks rather than attacking ones, but the trend toward spectacle and the shivery pleasures of terror prevails, particularly in J.M.W. Turner's 1840 masterpiece *The Slave Ship*, originally titled *Slavers Throwing overboard the Dead and Dying – Typhoon Coming on*.

J.M.W. Turner was a master of the sublime and an important painter in any number of ways, all of them evident in *The Slave Ship*. We see the sublime first in the intensity of the spectacle and the intensified colours, then in the terror implicit in the approaching storm, a demonstration of nature's far superior power. Throughout his career, Turner was thrilled by the Romantic vision, emphasizing the awesome power of nature and the corrupting effects of industrialism. In this painting, the corrupting influence was based on an actual event, the jettisoning of 133 slaves aboard the slave ship *Zong* to ensure an insurance settlement, which could be collected on drowned or shark-eaten human cargo, but not on slaves who died from sickness. The typhoon was Turner's fictional addition. He was inspired to depict this outrageous 1781 event after reading *The History and Abolition of the Slave Trade*, an abolitionist book by Thomas Clarkson.

Sharks followed all ships, including slave ships, as these vessels trailed a rich range of meaty smells, but the frightening lore about sharks and slavers was sensationalized by the abolitionists' campaign prior to the British banning of slavery in 1833. As William Boelhower writes, 'The slave ship and the shark are connected in popular and historical memory largely because of the effects of those who sought to abolish the slave trade. John Wesley, the founder of Methodism, wrote about sharks in his famous essay "Thoughts Upon Slavery," published in 1774' (Boelhower, 2013:132). Propaganda for a worthy cause may be an honourable act, and there is no doubt that the abolition of the slave trade was long overdue. But promoting terror of sharks was the collateral damage. As Boelhower notes, Wesley's and other abolitionists' efforts to propagandize vividly 'made the shark an important character in the drama of life on the slave ship, indeed a symbol of the violence and terror of the trade'. Referring specifically to Turner's painting *The Slave Ship*, Boelhower contends that 'the image of the predatory shark in the slave trade reached its apotheosis in the famous painting of 1840 by J.M.W. Turner' (Boelhower, 2013:132).

Scrutinising Turner's painting, one wonders what kind of 'apotheosis', or culmination, depicts the predatory shark as a sea monster with a fanciful snout (or, in other spots on the canvas, some sort of voracious sea bass)? Was Turner so ignorant about sharks' actual appearance or just indifferent to portraying them accurately? The answer is complicated by his intentions to evoke abolitionist outrage while also depicting the sublime, in this instance the awe-inspiring power of the sea. Realism was a secondary, or even tertiary, concern. Indeed, many of Turner's later paintings appear to be works of frenzied brushstrokes with an impressionistic effect, and very little precise detail, in his desire to capture the drama of the moment. *The Slave Ship* is more a study of mood than of historical event. At first glance we discern little more than a ship stripped of sails veering away from some commotion in the sea. The billowing turbulence of air and water creates a murk, shot through with vivid, unexpected colours. Only as we peer closely do we make out the human limbs and then the fish in the churning water. Are these fish biting into the victims? The excited seagulls circling overhead suggest that there may already be scraps of meat in the water. As Franny Moyle describes the scene, 'it is the pathetic hands and manacled feet grasping vainly through the foam, and the frenzy of fish rising to the surface in anticipation of their imminent meal, that convey Turner's sense of the tragic obscenity of the event' (Moyle, 2016:393). And yet within this gloomy scene and obscurity are bright flashes of light, suggesting lightning but also injecting unusual, even hypernatural colours. We may struggle to comprehend the painting's events, but we don't fail to recognize its powerful drama. Edmund Burke observed that 'the passion caused the great and sublime in nature, when caused to operate most powerfully is astonishment; and astonishment is that state of the soul, in which all its motions are suspended, with some degree of horror' (Burke, 2009:51). The sensationalizing of sharks was Turner's deliberate choice to provoke a passionate emotional response to his subject.

Shark in the Gulf Stream

When viewers are asked to recall Winslow Homer's famous painting *Gulf Stream*, they most vividly remember the sharks at the bottom of the frame (Shaw, 1997:81) – not the huge swells, not the black sailor precipitously tilted on a de-masted sailboat; not the angry sky or the waterspout in the background or the ship that could save the sailor but seems to be sailing away (Figure 5.4). Powerful or threatening as the painting's other central and surrounding elements may be, we focus first and most memorably on the sharks lolling on their sides as they circle. They are not biting anything or acting menacingly. They're completely accurate in shape, colouring, fin placement, neither exaggerated nor sensationalized. And yet they create the spectacle. If we coolly consider the situation and probable outcome, we recognize that like all shipwrecked sailors this sad fellow is more likely to die of dehydration than of shark-bite. And yet – the sailor is black and the sharks are big. The lore of the slave ships instructs us that these sharks don't need to attack, they will be provided for. Homer's *Gulf Stream* captures a moment between disasters: the storm that de-masted the schooner and the anticipated disaster of the sailor's body falling into the jaws of those sharks. One detail that is rarely noticed is the sugarcane poking out of the boat's hold. Sugarcane was commonly transported on small sailing sloops and schooners along the Gulf Coast, particularly from Cuba to Florida, and Homer was well aware of that (Wood, 2004:81). But the conjoining of sugarcane with a black man in the Caribbean yields a meaning that is unmistakable. As a plantation crop from Louisiana to Florida and throughout the Caribbean, sugarcane was more dependent on slavery than even cotton. *Gulf Stream* was completed in 1899, well after the abolition of slavery in the U.S., so we may wonder why the issue of black enslavement would still apply. The answer can be discerned in the sailor's posture of despair and the angle of the ship in the background, tacking away rather than offering hope of rescue. The black sailor is abandoned, as well as lost. Homer was highly sensitive to the plight of black Americans in the segregationist, Jim Crow South (Shaw, 1997:30–1). Lynching and other atrocities had become common. Alain Locke, the great Harlem Renaissance historian, has argued that Homer's painting is the first instance in which a black man is the central focus in American art (Shaw, 1997:29).

Homer sketched and painted several studies prior to *Gulf Stream.* One of these studies depicts a derelict boat with three sharks, no sailor or approaching storm, and a relatively calm sea. A second study shows a wider sloop, also de-masted, and a black sailor wearing a hat and shirt. This incomplete painting includes only one lolling shark, no waterspout and no sugarcane in the boat's hold. The sailor looks less despondent and less precipitously positioned on the sloping deck. *Gulf Stream* is thus an advancement in terms of its drama and its accumulation of resonances beyond realism. Within the sequence of paintings leading up to *Gulf Stream*, we recognize Homer's earlier New England works as gorgeous and emotional, but also overlookable without the context of his more overtly ambitious and mythic paintings, such as *Undertow* and *Shark Fishing* (both 1885) and *Gulf Stream* (1899). Homer reaches for wider relevance and wider audiences. *Gulf Stream*'s resemblance to Turner's famous *Slave Ship* was not lost on the critics of the day (Shaw, 1997:81), and such recognition certainly mattered to Homer as an artist striving to make a reputation and a living.

Photography

By the mid-twentieth century, thanks in large part to advancements in underwater breathing gear, photography became the dominant art form for depicting sharks, both to inform us and to frighten us. The pioneer in underwater photography was the Austrian Hans Hass, who made his first underwater documentary *Pirsch unter Wasser (Stalking Underwater)* in 1940 and went on to produce twenty-five books of photographs, many of them featuring sharks (Ellis, 2012:41–2). The most famous diver and underwater photographer was, of course, Jacques Cousteau. As Richard Ellis, himself an eminent underwater photographer, observes in his book *Sharks: A Visual History*, 'it was Jacques-Yves Cousteau (1910–1997) who popularized recreational diving and underwater photography as no person before him', having invented the Aqua-Lung with Emile Gagnan in 1943. The adventure of that invention, along with photographs of encounters with open ocean sharks, was detailed in Cousteau's hugely influential book *The Silent World* (1953), which sold half a million copies (Ellis, 2012:42). Among other popular books of underwater photography, Cousteau, working with his son Philippe, published *The Shark: Splendid Savage of the Sea* (1970), which presented sharks as 'savage denizens of the deep'. Such dramatic prose and photos must have smelled like blood in the water for the editors of *National Geographic* magazine, which eagerly reproduced Cousteau's work and commissioned other underwater photographers, such as Jerry Greenberg, to produce photography books like *The Fish Men Fear . . . Shark!* and the photos for the 1968 *National Geographic* cover story 'Sharks: Wolves of the Sea' (Ellis, 2012:44–5). The unfortunate effect of all this colourful photography was less to educate or inform than to entertain with fright, and to sell a lot of books and magazines.

For capturing terror in photographs, the champions have been the Australians: Rodney Fox, who was nearly killed by a white shark in Spencer Gulf, and Ron and Valerie Taylor, who captured the iconic images of huge, open-mouthed white sharks displaying pure power and propulsion with eyes so black they look like hollow holes. These awe-inspiring photos epitomized our shark fears, bridging the gap between Copley's shark sublime and Steven Spielberg's film *Jaws* and inspiring nightmares in their own right. The difference is an important one. Instead of working safely on land, Valerie and Ron Taylor participated in the long and arduous production for the documentary *Blue Water, White Death*, which required that they jump into the Indian Ocean beside a dead whale to film the slow-motion feeding of some of the largest predatory sharks.

The Australian Paul Hilton's work is hazardous in another way, documenting the enormity of shark-finning, which kills an estimated 100 million sharks per year and has pushed some shark species to the edge of extinction. From his home in Hong Kong, Hilton travels across Southeast Asia, the market for shark fins, capturing in poignant images the reality behind the numbers of sharks destroyed. The German photographer Thomas Peschak also documents the brutality and enormous waste of the shark-fin trade, with unblinking images of blood-splattered carnage and industrial-scale killing, with a photo of men dumping bags that each contain hundreds of shark fins, and a night shot of more than a hundred shark carcasses neatly positioned and arrayed in six long lines across the pavement beside the docks. Peschak's photographs of living sharks are gorgeous, showing sharks circling in the water column or feeding on anchovies. He is perhaps best known as the *National Geographic* photographer who gave us the image of a white shark closely following a kayak, an image less frightening than beautiful since the water is a clear, iridescent blue and the man in the kayak is glancing back with an expression more comical than panicky. As Peschak has said, 'Rather than adding to the growing stock of menacing, toothy white shark photographs, I was driven to capture the gentler side and prehistoric perfection of this animal' (http://time.com). This kind of photography is welcome amid ongoing international efforts to change the image of sharks through thoughtful and responsible representation.

The list of talented shark photographers is too long to be comprehensive here, but a few others should be mentioned, if only for the range of their depictions. David Doubilet has shot 60 stories for *National Geographic*, including the cover image of the enormous white shark with equally enormous toothy jaws for the April 2000 shark issue and the flying white sharks (popularly known as 'Air Jaws') of False Bay, South Africa. But these are just his most famous images. He has also photographed hundreds of other sharks, including bull and tiger sharks, whale sharks and basking sharks, species that inspire our awe. The contemporary photographer Jeff Rotman has said that 'in my business, sharks are box office' (http://time. com), and yet he specializes in capturing images of sharks that show no teeth. His close-up of a new-born spiny dogfish, still attached to its umbilical sac, is a celebration of sharks' beauty, as is his shot of a hammerhead's gorgeous blue eye. Other shots, like Rotman's very tight close-up of another shark's nictitating eye, demonstrate better than words one of nature's most ingenious biological solutions, the membrane that automatically shutters the shark's eye in the final moment before it bites. Similarly, Rotman's photos of odd-looking shark species, such as the wobbegong and Port Jackson sharks of Australia, display the delightful variety of sharks (of which there are 500 species) as natural experimentations. As a lover of sharks, Rotman also documents the ugliness and waste of shark-netting and shark-finning, showing, for example, a tiger shark that has been drowned by a shark net in Durban and a large group of finless dead sharks lying on the ocean floor off Costa Rica. What these photographers offer us is not sharks sensationalised as potential man-eating machines, but their natural beauty and grace. They show us the glorious rather than the gloomy sublime. They reveal the majesty and elegance of these much-maligned creatures.

A range of response in twentieth-century art

Phillip Evergood began as a social realist before veering into fantasy and symbolism. He painted people, not fish, yet his *The Rubber Raft* (1945) is extraordinary, reflecting the widespread fear of sharks that was shared by American sailors and airmen serving in the Pacific in the Second World War. This intensely colourful oil painting shows two ragged castaways in a pathetically small yellow raft surrounded by at least a dozen sharks, many of them lolling on their sides like Winslow Homer's sharks, all of them featuring lips redder than those on Copley's shark. By referencing Copley and Homer, the painting recalls the origins of our shark terror, in the sensational sublime, with sharks stealing the scene. And yet these sharks seem cheerful, as are the colours, the bright blue and the shades of yellow and orange and violet around the sun. The painting resembles a child's happy daubing, the castaways' likely deaths notwithstanding.

Another notable artist is Richard Ellis, who paints and draws marine creatures with great accuracy. His sharks are genuine portraits, neither fanciful nor hyperbolic. With his background in biology and natural history, Ellis is able to depict a wide range of shark species with extraordinary emotional emphasis as well as anatomical accuracy. His painting *Grey Nurse* (1975), for instance, captures a solitary shark in the dappled light from the surface, with the patterns of illumination creating a similar pattern on both the shark and the sandy ocean floor, uniting the elements as well as accentuating the distinctive shape of the shark and its shadow on the sand. More than a hundred of Ellis's paintings were selected for a travelling exhibit by the Smithsonian, and they are now in the permanent collection at Whaleworld Museum in Albany, Western Australia.

The next major landmark in the depiction of sharks in art is, unavoidably, *Jaws*. Roger Kastel's painting, which became the cover of the novel's paperback edition, is itself iconic, providing the image for the famous movie poster (Figure 5.5). The painting shows a young woman (Chrissy in the film) swimming an adept crawl on the surface while, unbeknownst to her, the head of a spiky-toothed shark the size of a prehistoric megalodon is propelled up toward her. But films are more socially powerful than any paintings. No artwork has done more damage to sharks' reputation than *Jaws*.

The movie takes pains to reference historical incidents of shark behaviour. For instance, Brody, the aquaphobic police chief, studies a book on sharks that we plainly see features the *National Geographic* article 'Sharks: Wolves of the Sea', and which has the Canadian artist Paul Calle's painting on the cover. Calle based his dramatic scene on a 1953 incident in which a twelve-foot white shark capsized a lobster boat off Cape Breton Island, Nova Scotia. In fact, the lobstermen were rescued unscathed, but in the painting the shark stoves in the boat hull with its huge snout and the lobstermen are sent reeling. Brody glimpses Calle's painting, charges to the window, and screams at his kids to 'Get out of the boat!' It is a comic moment, tapping into popular fears based on misrepresentations and exaggerations like Calle's or indeed the film's, while mocking Brody's hysteria.

Damien Hirst and beyond

While it may be true that *Jaws* was disastrous for sharks, Spielberg and Benchley did not commission their deaths, unlike Damien Hirst. *The Physical Impossibility of Death in the Mind of Someone Living* (1991) was sold for $8 million in 1995; Hirst then sold a second and much smaller shark to a South Korean dealer for $5.7 million. After the first pickled shark began to rot in its vitrine, Hirst replaced it with another tiger shark, whose killing he had arranged with the notorious Australian shark hunter Vic Hislop.

The scale of Hirst's work is impressive. His vitrines are twenty feet long. To prepare the replacement shark, roughly 228 gallons of formaldehyde were used for the operation. The artist and his assistants worked in shifts to minimize exposure to the toxic fumes. Hirst went on in 2010 to create further shark works, one with a white shark, ironically entitled 'The Immortal,' and another with a hammerhead, which he entitled 'Fear of Flying' based on the resemblance between the shape of a hammerhead's cephalofoil and that of an airliner. Hirst is a very deliberate artist; he intends to create sensation. In his interview book, *On the Way to Work*, Hirst admits, 'I like creating emotions scientifically' (Burn and Hirst, 2001:13). Hirst's literally realistic sculpture is suspended with its mouth open wide, the better to show off its distinctive tiger shark teeth. Its eyes are open. We know that the shark is dead, but also that it is real. Indeed, the shark looks alive and, given its posture and sheer bulk, dangerous. As art historian Sarah Kent has written, 'The tension of the piece comes from its neutrality; from raising issues, yet refusing argument. The work offers drama without catharsis, confrontation without resolution, and provocation without redress' (Kent, 37:1994).

The interest surrounding Hirst's shark cannot be separated from its title (or from the artwork's outrageous price). The words *The Physical Impossibility of Death in the Mind of Someone Living* challenge us first to decide whether we can imagine death, although as living creatures we have to experience it; then we have to wrestle with the concept of *physical* impossibility, suggesting an inborn or instinctive inability, not just a psychological block. Hirst's title begs the question of whether sharks can be equated with death, not only because the shark is so dead but because it conjures our own deaths in some primal way. And is our response truly primal – that is, instinctive – or is it conditioned by hundreds of years of sharks' bad press and sensationalizing? In our imaginations, manifested in nightmares, films, photography, and fine arts, sharks hold a place of dread, even mortal terror. Yet those who dive with sharks, even the 'most fearsome' – the white shark, the tiger shark, the bull shark, the species most associated with attacks on humans – report a very different response. When we encounter them in the wild, sharks are shy, curious, often very tolerant of divers, majestic and truly awe-inspiring, rather than scary.

Hirst's famous sharks may harken back to Copley and the Romantic sublime, but they also reflect his solemn form of irony that speaks to 'now'. Art critic Jonathan Jones lucidly explains this intersection:

> Death, decay, the sublime were the themes of the British art that defined the end of the twentieth century; the horror of the shark swimming towards you through formaldehyde . . . The sublime was the aesthetic of these years, this art – and the sublime, as the eighteenth century politician and thinker Edmund Burke argued, is about power. The origin of the sublime, for him (Burke, not Hirst), lies in our awe before a majestic, divine authority. Today it is an awe of art itself, or at least a desire to experience that awe; to be knocked over by art, to be kicked in the teeth.
>
> (http://arts.guardian.uk)

Damien Hirst's posthumous sharks are, of course, post-*Jaws*. We couldn't respond to them as powerfully if not for the phenomenon of *Jaws* at the apex of the sensationalizing (and distorting) sublime. Charles Saatchi commissioned the work and subsequently sold it to a billionaire hedge fund investor, who explained, 'I grew up in the generation of *Jaws*. I knew it was *the* piece of the 90s' (www.nytimes). But just as *Jaws* caused a seismic response, Hirst's *Physical Impossibility of Death in the Mind of Someone Living* provoked a counter wave. The Irish artist Finn Stone has made a series entitled 'Great White Artist' riffing on Hirst with a series of sculptures of cheerfully polka-dotted sharks (all of them fibreglass, none actual pickled sharks). Other artists also aestheticize the shark, particularly its design. Xavier Veilham's 2008 stainless steel sculpture *The Shark*, exhibited at Galerie Emmanuel Perrotin, Paris and Miami, depicts a sleek, five-metre shark in planes of seamless stainless steel, the better to celebrate its beautiful, elemental shape. The gallery label explains:

> This great five-meter shark echoes a powerful collective imagination . . . a prehistoric animal, presented on a device recalling the supports of the

> skeletons of the dinosaurs in the museums of natural history. Like Le Lion, it is a fantasy, disturbing and indomitable creature, evoking the 'lines-forces' of the plastic dynamism theorized by the futurist painter Umberto Boccioni: simultaneous perception of movement allied to the translation of speed. If the work recalls Brancusi's marble fish for its elegance and Damian Hirst's formal beast by its derision, it mostly appears as Stealth's underwater alter ego.
>
> (perrotin.com)

In response to *Jaws*, to Damien Hirst and other sensationalists going back to John Singleton Copley, artists have used wit to comment on the exploitations of our fears. And others celebrate sharks' beauty without resorting to evocations of fear or depictions of teeth, evincing awe and respect for the power of sharks, emotions that underscore reverence and appreciation, and counter the western desire to annihilate our competition. One post-*Jaws* (but pre-Hirst) response is the Headington Shark, a life-sized sculpture of a white shark crashing headfirst through the roof of an ordinary townhouse in Oxford. Created in 1986 to mark the anniversary of the Nagasaki atomic bomb attack, the Headington Shark is complicated in its protest mockery, apparently linking the bomb and the Spielberg film as violent intrusions into our peaceful domestic life, events that shock us every time we regard them.

The artist David Showalter offers a gentler corrective to Hirst's media sensation. He has painted a reflective series of acrylic depictions of grey reef sharks and their geometry, showing them in empty spaces and orbital patterns, featuring the distinctive shadows that they cast. As he explains, 'After getting past their reputation, I was struck by their elegant lines and geometric beauty' (http://davidshowalterart.com). 'Reputation' is a word we use for the general opinion of others. As an artist and an independent observer, Showalter was able to view sharks in calmer and more accurate terms.

Several contemporary sculptors construct their sharks from consumer goods to make material an ironic point. A self-described 'junker', Robbie Barber used discarded golf bags to create the seven sharks in his work *The Reef* (1992), with oil drums and other refuse forming the reef itself. Kitty Wells salvaged parts of such old appliances as a Maytag refrigerator and a Hotpoint stove to create the circling sharks that she hung from trees in *Pine Sharks* (1997). And Vancouver artist Brian Jurgen repurposed Samsonite luggage to remind us of the resemblance between sharks and airliners in his sculpture *Crux* (2008) (http://www.smithsonianmag.com).

Graphic sharks

Shark cartoons are by definition ironic, lampooning their subjects and usually portraying sharks that talk. These depictions are never about sharks themselves but their place in the popular imagination – or about how sharks would behave if they were us. They're also, often, quite funny, mocking our sensational fears as well as our own sharkiness. The lawyer as shark is one of the most frequent tropes, with one cartoon featuring a businessman shaking the hand of a sly-looking fellow with slicked-back hair and an obvious dorsal fin bulge on the back of his suit jacket. The cigar-puffing businessman is saying, '. . . *as soon as I saw you I knew you were the right man for the job*' (cartoonstock.com). Another cartoon shows an early amphibian crawling out of the sea, closely followed by a shark who is saying, 'Trust me, you're going to be very happy you brought along legal representation' (i.pinimg.com/new-yorker.jpg).

Other tropes mock our own tendencies to eat sharkily, with one shark remarking to another in a Leo Cullon cartoon, 'I usually have a light breakfast, then a feeding frenzy in the early afternoon, and that's it for the day' (i.pinimg.com/originals). In another cartoon, one white shark is leaning close to another's head, saying, 'You've got a thong caught between your front teeth' (allposters.com/thong). In a similar vein, two sharks are swimming toward a beach, with one remarking, 'God, I love the smell of sunblock in the morning' (i.pinimg.com/236x/37/af/5d/37af5d b5843fdfe1cbe460b467e520a3.jpg). A very recent cartoon shows Donald Trump sitting behind his desk with his characteristic scowl, gazing at an array of unpopular animals in front of him: a vulture, a tarantula, a scorpion, a rattlesnake, a rat, a hyena, a warthog, and a particularly nasty-looking shark in a giant fish bowl. One of the president's aides is explaining, 'A pet always boosts a president's approval rating' (art.com). The reputation of sharks suffers from such comparisons.

Most shark cartoons simply satirize the dopiness of humans. Using the mannequin trope, one cartoon shows a white shark with a mannequin in its mouth explaining to a second shark, 'The doctor said it might help me quit' (today.duke.edu). A Larson cartoon shows crates marked Acme Mannequin Co. spilling off a sinking ship, with one large shark saying to another, 'What is this? . . . Some kind of cruel hoax?' (i.pinimg.com/Larson). In a panel by J. Longo, one shark between bites is saying to the other, 'I thought being a humanitarian meant only eating humans . . .' (static1.squarespace.com). The desert-island trope also inspires shark jokes that satirize the human tendency to act stupidly, with a pair of castaways looking across a shark-infested stretch of water at the Tiki bar on the next desert island and one saying to the other, 'Keep in mind it was put there by the sharks' (condenaststore.com).

Other tropes are more complicated. A *New Yorker* cartoon shows a lifeguard stand with a sign on it saying 'Lifeguard On Acid' and the lifeguard on his feet in great alarm hollering, 'Sharks! Hundreds of them! Flying sharks! Everyone out of the water! They're converging on the future!' (media.newyorker.com/cartoons). The interpretation of many viewers today would be that this joke is based on the wacky *Sharknado* films, but in fact the cartoon was prescient, predating these campy movies by more than two years.

How do we explain the persistent popularity of shark cartoons? The use of tropes offers a partial answer: the familiarity of recurrent situations launches the joke, allowing us to catch it faster; we see desert islands, lifeguard stands, or large sharks talking underwater, and we're primed with certain absurd, ready expectations. Furthermore, our shark fears prepare us for nervous laughter. And, generally, we love talking animals that exhibit our dumbest or most recognizably reprehensible behaviours in exaggerated form. The reason behind this last explanation is one of our best traits as a species, our easy willingness to laugh at ourselves, which bespeaks a modesty, a recognition that we're not necessarily the best or the most powerful creatures on earth.

Movie posters apart, other graphic shark representations, such as those in advertising and promotional posters, celebrate sharks for their streamlined design and their beauty. A BMW ad shows a sleek shark shape, with the car's triangular roof antenna exaggerated to become a dorsal fin and the four side vents suggesting shark gills. The whole design is raked into an elegant angle that portrays streamlined propulsion.

The Arena Water Instinct Company, promoting 'powerskin' high-tech swimwear that uses a scaled fabric inspired by sharkskin, creates a similarly positive association with the elegant design of sharks, showing a springboard diver's perfect pair of legs and perfect splash with a mako's equally shapely torso connected to the legs beneath the surface of the water. A 1938 poster produced in the US by the Depression-era Works Progress Association takes the familiar anti-shark tack, filling the vertical length of the poster with a sinuous, scary shark and offering the words 'Pneumonia Strikes Like a Man Eating Shark Led By Its Pilot Fish The Common Cold,' thereby exploiting shark fear for public-health purposes.

Sensational shark depictions, from the sixteenth-century drawings of grotesque, desiccated sharks to Damien Hirst's pickled sharks, raise questions about why we're so fascinated by animals that, on rare occasions, might kill and even eat us. Are we obsessed with them because they puncture our monstrous arrogance, specifically the belief, enshrined in the Old Testament, that we enjoy the God-given right to exploit all the other species of the earth? (Genesis, 9:2). Except for their fins, sharks are not our favourite food, but like wolves, bears, and coyotes, they represent competition and thus challenge our dominion. Mere rivalry does not account for our fascination, however; such a deep obsession must be psychological or religious in origin.

The work of the contemporary artist Richard Longo provides an interesting answer. His series *Perfect Gods* (2007) features massive hyper-realistic drawings of white sharks in their fiercest poses and in close-up. The backgrounds are a velvety black created by thick-textured charcoal, so that the sharks appear to lunge out of deepest darkness with an equal darkness extending down the shark's throat. The shark heads, teeth, and torsos are photographically accurate, but these portraits are actually drawn, their medium described as 'archival pigment' or 'charcoal mounted on paper'. The rows of teeth, with the huge triangular teeth on top and the rows of curved sabres on the bottom are perfectly rendered. The thought one has while gazing at these portraits is not 'perfect gods' but 'perfect monsters' – except, of course, that gods may be monsters, which is a central part of the definition of the sublime.

Many of Longo's shark paintings measure a metre square, and one, entitled *Triptych*, actually measures 2.77 metres by 4.87 metres (four to five metres being the average size of a young adult white shark). In this huge painting, the front half of a single white shark is displayed from the underside, with not just the whiteness portrayed but also the rippling musculature that undulates as the shark swallows. There is no escaping the fear that this painting evokes of overwhelming, beyond-our-imagining power. The thick blackness reinforces these associations with a satanic evil. What Longo offers us, particularly in conjunction with the series' title, is not merely a mortal terror, but an immortal source of terror: evil incarnate, the demon that emerges from the blackest gloom. He offers us the epitome of the darkest definition of the sublime, even as he critiques our enchantment with it.

Art in Oceania

Artists and artisans in Polynesia, Micronesia and Melanesia never suffered from such endemic sensationalism, or from scriptural arrogance, or from the self-mockery that naturally follows such excesses. Pacific shark artists display an awe without terror, a deep respect and often a reverence for sharks that fulfil the other side of the sublime, recognizing the glory without the gloom. Sharks are, for Pacific islanders, simply part of the donnée. Large or small, sharks are in the water with them all the time. Visual associations that would evoke sharks' ferocity and power are not overlooked in Pacific art, but neither are they featured, much less stressed. Across the Pacific, from Mexico to Australia and Asia, sharks occupy what may seem to European or American eyes a remarkably modest place.

Our differences are many and profound. As the eminent ethnographer Adrienne Kaeppler has pointed out, 'Art as a separate category has little meaning in an analysis of traditional Polynesian culture. Polynesian languages do not have indigenous words or concepts for 'art' (Kaeppler, 2008:7). Kaeppler does not claim that Oceanic cultures have no art, nor that the motives underlying art everywhere else in human society do not apply to the Pacific Islands; only that art and artist are undistinguished from other regular activities. Making art is normalized. Levels of skill and talent or innate ability have always been recognized, and the work of talented carvers, embroiderers, and creators of ceremonial weapons and objects celebrated. However, the category of artist wasn't elevated, like that of a sage or priest, and the ceremonial objects were not housed in a museum or curated. Pacific Island cultures were more utilitarian than European, American, or mainland Asian cultures, since raw materials, including the ores that produce metals, were scarce or simply not available. Until relatively recently, islanders had few imports and had to make full use of what little resources they had. Shark teeth, prized for their sharpness, were used for weapons and cutting tools; sharkskin was used for its rough texture and tensile strength.

Island societies are necessarily more cooperative and less competitive within their own societies, but it is facile and condescending to romanticize the customs of others. Pacific Islanders were not pacific. They waged war regularly. The more peaceful people, such as the Chatham Islanders of New Zealand, were conquered by their more bellicose neighbours. We find shark teeth used in ornamental weaponry from Hawaii to New Zealand and throughout Micronesia, and not merely to suggest the mystical power of the shark – also to maim and terrify the enemy. A knife collected on Captain Cook's third voyage to the Pacific featured a white shark's large triangle tooth lashed onto the tip of a sturdy, curved wooden handle (Hurst, 2006:115). For greater effectiveness, shark-tooth weapons were sometimes pegged or embedded with wooden splints fit into the grooves of clubs or battering paddles; other times, holes were drilled and the teeth lashed to the paddle's edge (Hurst, 2006:82). Such weapons would not merely pierce but maul one's enemies by raking open their skin. 'Knuckle-busters' were typically studded with sharks' teeth.

A *leiomano* was a shark-tooth club used all across Polynesia and especially in Hawaii, where the word translates literally as '*lei* of shark', *mano* meaning shark. The use of shark's teeth was not just practical – their teeth being the sharpest, hardest, pointiest things around – but psychological (Figure 5.6). The Kiribati people of the Solomon Islands made spears wrapped with rough shark skin by lashing three wooden prongs together, each edged with small shark teeth and embedded with larger ones (Kaeppler, 2008:135). The art historian Christraud Geary raises an interesting point when he remarks that 'This delicacy of handiwork contrasts with the intimidating nature of the object' (Geary, 2006:135). But does delicate artistry contrast with violence and intimidation? We may recall medieval weapons and helmets. The time and talent required for such detailed and elaborate handiwork transmits a visual message about the power and importance of the warrior employing that weapon. Sharks' renowned power and ferocity are also implicit in their representations in Pacific weaponry, except not in the Western sense that implies that the warrior has killed sharks to make his weapons. In Papua New Guinea, for instance, ceremonial masks and shields featured triangle designs based on the triangular fins and teeth of sharks, to establish spiritual and totemic associations.

Beyond the utilitarian purposes and weaponry, shark parts were used in ornamentation all across the Pacific, mainly for men. Highly ornamented Tahitian gorgets, or neck girdles, originated as throat protectors for warfare, but evolved into strictly ornamental wear. These gorgets became quite elaborate, featuring shark teeth along with feathers, pearl shells, wicker, and often a dog-hair fringe (Kaeppler, 2008:12). The Nauru people of Micronesia also created ceremonial girdles out of sharks' teeth, along with other decorative objects such as frigate bird feathers, coral, and shells (Kaeppler, 2008:23). In these instances, shark teeth carried little or no psychological or spiritual significance and occupied no focal place on the girdle as a piece of body art; shark teeth were simply plentiful among the shells on the beaches, since sharks were plentiful and constantly shedding their teeth. In the Society Islands, we find ceremonial dance costumes, headdresses, and pendants that sported rare shells rather than the less special shark teeth. The Maori, too, preferred pearls or greenstone as costume elements since sharks' teeth were neither precious nor rare.

However, in tattoos – body art – the triangular shark symbol carried a distinct spiritual significance, associating the tattooed person with powerful spirits, sometimes seen as gods and sometimes as ancestors. After a number of generations, the active memory of an ancestor fades since there is no longer anyone alive who can remember that ancestor. But because the memory of a founding ancestor may be an important organizing force for the community, Polynesian people memorialize the 'life force' from which he (or she) came. If, for instance, a person's essential life force was believed to have been derived from the fierce and forceful shark, then after four or five generations his descendants might begin to remember him as a shark rather than as a human. In many Oceanic societies, a person's life force can take precedence over what we think of as 'personality' (Geary, 2006:13). Tattoos often recall and honour one's ancestors in symbolic, graphic forms, with the triangular shark tattoo expressing power, ferocity, and guidance. In Hawaii, extensive tattooing was common at the time of Captain Cook's voyages, both for expressions of identity and ancestral connection and, with a tattooed row of dots around the ankle, as a charm against sharks (Kaeppler, 2008:115). Extensive body art was also common among the Maori of New Zealand – and continues to be practised, since Maori culture has been less undermined than that of other Oceanic peoples. Their body art is often so elaborate that it looks like stencils, sometimes including the curving shapes of sharks. Tribal art today replicates these tattoos as actual stencils in carvings, prints and paintings, as well as in the shapes of earrings and necklaces.

Polynesians, who were the ancestors of the Maori, considered sharks to be guardian spirits. Hawaiian reverence for sharks is called *aumakua*, which means 'family guardians,' and often refers to particular local tiger sharks as the carriers of that ancestral guardianship. Maoris used shark teeth in jewellery, with white shark and mako teeth being the most prized. According to Maori mythology, the demigod Maui formed the Milky Way by placing the shark Te Mangoroa in the sky (teara.govt.nz). Maori shark tattoos and stencils look much like patterns of stars in the crowded night sky of the Southern Hemisphere. The shark totem also appears in Australian aboriginal art, reflecting its importance in Dreamtime, when it is believed that ancestral spirits emerged from the unformed earth, sea and sky, giving shape and meaning to the world. For instance, in Arnhem Land, in the north of Australia, the ancestral whaler shark *Mana* was believed to have formed the land in a rage, charging inland and slashing with his teeth to carve out river systems, after being speared by another ancestral spirit. Dreamtime symbols are the basis for most contemporary Aboriginal art.

The artists of the Torres Strait, in the northernmost part of Australia, are particularly successful as unifiers of different Pacific cultures, Aboriginal, Melanesian, even missionary Christian, by interpreting traditional stories and myths in prints (particularly linocuts) and in sculpture. Sculptor Ken Thaiday creates headdresses for dance, chant and drumming ceremonies using the same materials that Polynesians and Melanesians employ in their art: feathers, hair, bamboo, and shells. Although much influenced by Australian Aboriginals, the people of the Torres Strait are in fact Melanesian. Thaiday aligns himself with both traditions and belief systems, claiming the hammerhead shark as his totem and portraying it in his art (australianmuseum.net.au). His sculptures originate not just from ancestral traditions and beliefs, but also from his own close observations of the natural world. As he says in an interview with the Australian Museum at the time of their purchase of his *Triple Beizam Hammerhead Shark Headdress*:

> Look at the bamboo, the shark is camouflaged three times in the water. Brown, grey and blue. When they go in the water and turn the colour to blue . . . well you'll see they come up the next point of the reef, colour changing again. You see them like they come in brown, grey, that colour. That's one good thing about the shark. The front, all these two feathers, the same thing they represent the speed of the wind, the wind blow. Centre feather of the shark. When they come up, you watch them they come up like this [blows], they're very powerful.
>
> (australianmuseum.net.au)

Safina Stewart similarly unites different cultures, and her art similarly celebrates the vividness of the natural world. Although she now lives in Melbourne, Safina's heritage is both Aboriginal and Torres Strait Melanesian. Her non-indigenous heritage is Scottish, as her surname suggests. Her father was a missionary and inspired Safina's deep Christian faith. In both her prints and her paintings, she creates intricate designs that strongly suggest Aboriginal Dreamtime patterns. Although Aboriginal beliefs are polytheistic, they are based on belief in majestic origins. Safina's Duel Jewel (2007), an acrylic on canvas, shows two sharks that she describes as 'majestic and powerful and hunting in tandem to 'remind us of their powerfully majestic Creator and the strength given when working well as a focused team' (http://indigenousjesus.blogspot).

We find intriguingly similar tendencies all the way across the Pacific, among the indigenous people of Mexico, particularly in the Zapotec culture of the Juchita isthmus near Oaxaca. In her photograph 'Tehuantepec' (1980), the Oaxacan artist Graciela Iturbide stages a scene with a playfully smiling Indian boy wearing a set of shark jaws around his head. Behind him is a mural depicting a large wading bird, likely an ibis, whose pointed beak seems to touch the shark jaws. Iturbide has stated her intention to depict the 'complicity' between photographer and subject (jordanelgrably blog), and the boy's playful smile invites us into this 'complicity'. But into what exactly? We recognize the bird's suggestion of mythology and the boy's indigenous ancestry, but what about those shark jaws that touch or at least align with the beak of the bird? The art critic and scholar Jordan Elgrably writes that 'animals are an important part of Zapotec culture, as witnessed in many of Iturbide's photographs. Referring to themselves as *Vinigulasa* – "the people of the clouds" – Zapotec legend has it that man descended from the heavens in the shape of birds of incredible beauty.' He also notes that the ancient Zapotec custom of finding an unborn child's *nahual*, or alter ego, which consists of assigning the baby an animal representation, continues to this day. As Graciela Iturbide explains, 'When a woman is pregnant and about to give birth, people come to her home and draw animals in the dirt, which they continually erase. The animal drawing left in the dirt at the time of the infant's birth will be his *nahual*' (jordanelgrably blog).

Returning to the visual alignment of the bird's beak and the shark's jaws, we may now reasonably conclude that one mythological spirit is connected to the other, as the bird appears to be whispering some secret to the shark.

In eastern Polynesia, the use of sharkskin membranes on drums, also known as membranophones, was widespread (Kaeppler, 2008:53). Such membranes had an obvious practical value since sharkskin is tougher than other materials that Polynesians had access to. But given the spiritual importance of shark teeth, both as actual objects in ceremonial costumes and as symbolic shapes in carving, tattoos and painting, we can surmise that sharkskin was employed for more than mere utility. Formed by dermal denticles, a form of tooth, sharkskin has an unusually rough texture and appearance; it offers a sensory and aesthetic appeal. Sharkskin, known as shagreen, has long been a valuable material for decorative jewellery boxes and fine specialty cases in Europe and America. Shagreen was expensive since working out the denticles was laborious. Beginning as a guild craft in Holland in the eighteenth century, shagreen achieved the height of its importance as an art-form material with John Paul Cooper, a former architect of the Arts and Crafts movement, whose London studio produced all manner of ornamental objects, including vases, candlesticks, boxes, and picture frames made of sharkskin (Kuzmanovic, 1995:155). Buyers appreciated the beauty, the texture, the play of light on the surface of the material, and of course the prestige associated with the expense of shagreen, but they were not attracted by any spiritual connections to sharks or the power of sharks. Many buyers didn't even realize that shagreen was sharkskin. In Japan, however, the owners of samurai swords with sharkskin hafts surely recognized that the handle of the sword held associations beyond the advantages of a good grip and a beautiful appearance. For Eastern and Oceanic peoples, spiritual meaning and sensory appeal combined to form an aesthetic of beauty, religious significance, and utility – which is a full definition of art, even if 'artist' was not an occupation in the Polynesian world.

The crux, the defining difference between Western dread and Eastern respect or reverence for sharks, may be reduced to how we feel about being prey, seeing ourselves not at the top of the food chain but at a lower-than-alpha position. In the Western tradition, as well as in the Bible, our species asserts its dominion over the animals, as well as over all other forms of life. As David Quammen writes in his excellent book *Monster of God*:

> Have crocodiles and sharks committed ugly, horrific acts of homicide and anthropophagy? Yes, and by doing so they're offered us a certain perspective. While we humans may be the reflective members of the natural world, we're not (in my view, anyway) its divinely appointed proprietors. Nor are we the culmination of evolution, except in the sense that there has never been another species so bizarrely ingenious that it could create both iambic pentameter and plutonium. Throughout the course of the human story, one reminder of our earthly status has been that, at times, in some landscapes, we have served as an intermediate link in the food chain.
>
> (Quammen, 2003:13)

Being an intermediate – at least not the highest – link in the food chain is something that Pacific Islanders know all about. As Hawaiians express it, the tiger shark 'takes' a swimmer or a spear fisherman. Micronesians who capsize in the open ocean or far out to sea do not expect to die slowly of hypothermia or to drown. Sharks big enough to eat us are simply part of the given, like typhoons and earthquakes and volcanic eruptions. The power of these forces is not only overwhelming but also random, beyond our control, and therefore can only be respected or revered, a word that we could define as respect.

The differences between Western culture and Oceanic civilization rest on greater differences than just our basic documents and beliefs. Industrialization, technological advancements, and of course geographical isolation are also determining factors. With economic and scientific progress comes control over our circumstances and some faith in our protection from outside, outsized forces, including natural disasters and extreme weather. The widely held wisdom among Polynesians and Micronesians is that after a tiger shark has been sighted near the island, spear fishermen should wait three days before venturing back in the water. Surfers from California to Western Australia, and notably those who have been bitten by sharks, speak humbly of their place in the sharks' domain. They accept the risks, without insurance or assurance. They share the pacific attitude of the Pacific Islanders.

The contemporary sculptor Andrea Everhart also demonstrates this maturation of attitudes, a union of opposites: fear of the awesome power of sharks and awe for their beauty. Her bronze sculpture of a white shark, entitled *Ocean Beauty*, is perfect in its details of lateral line, fins, nostrils, snout and teeth. It also includes elaborate patterning, strongly reminiscent of Polynesian tattoo art, across the shark's upper torso, dorsal fin, and tail. Describing her sculpture, Everhart writes, 'I find nothing more terrifying or beautiful than the great white. Its image has become what nightmares are made of, with its size sheer power, and rows of shark teeth set into its pink and roses [sic] flesh' (http://www.everhartfineart.com/).

Historically, the human view of sharks, as reflected in art, has been decidedly bifocal. In the Western tradition, we've mostly sensationalized, demonized, and then tried to extirpate whatever might challenge or eat us. The Pacific tradition is one of accommodation, accepting the sharks' role within the overwhelming power of natural forces. The trend that we find today, however, is (happily) a converging of attitudes. Even in the sensationalized West, we see a growing respect for sharks, demonstrated not only by our protection laws and conservation efforts, but also, even more visibly, by contemporary art. The exhibition 'On Sharks and Humanity', organized by Wild Aid with support from the Chinese government, featured Chinese and Western artists reconsidering our relationship with sharks and through them with nature. The exhibition opened in Monaco in 2014 and travelled to Moscow before its larger opening in Beijing at the National Museum of China. The exhibition 'On Sharks and Humanity' was designed not merely to aestheticize sharks or exploit their sensational appeal, but to be 'a platform for the education and humanitarianism' as the curator, Huang Du, asserts. Many of the works in the exhibit are strikingly beautiful. *Swimming* by Zou Liang is a full-sized white shark sculpture in dazzling silver acrylic. *Evolution* by Gao Xiaowu is also a silver shark sculpture but more expressionistic in form, with multi-coloured fins that seem to explode into wings. The painting *Us* by Liu Zining depicts a giant eye with an iris that is red-orange like the sun or the vertical pupil of a shark. Other works are more directly environmental. *Enlightenment* by Yu Yang is a shark-shaped sculpture made entirely out of spears or short harpoons. *The Body and the Shark* by Weng Fen recalls Damien Hirst's famous sculpture, except that Fen's vitrine contains living bodies, two white-tip reef sharks swimming briskly and one naked actual boy treading water very calmly. As the artist Fen Xueyi says in a video interview, the exhibit attempts to consider 'how we conduct ourselves' and 'to look back to the beauty of the original world' (wildaid.org).

We may wonder how the history of art, beginning with the career of John Singleton Copley, might have been altered if Captain Cook had focused on Polynesian art and ethnography instead of cartography during his three long voyages across the Pacific. The Pacific cultures that he encountered (and failed to successfully engage) believed in the acceptance of the natural world and in finding ways to revere those forces more powerful than humans, especially such overwhelming powers as volcanoes and sharks. As in the Western tradition, Polynesian shark art derived from a supernatural belief system, except one that veers toward reverence rather than toward evoking demons and terror. Necklaces and other art forms honoured and celebrated sharks, recognizing not only their power and possible divinity over us in their element, the ocean, but also their extraordinary evolutionary features like teeth and denticle skin. In ways unknown and opposite to Europeans, Polynesian peoples lived by a harmonic system of mythology and divinity based on a realistic and non-sensational recognition of sharks' place in the environment and of the environment's place in human affairs. This was the civilization that Captain Cook sailed into during his Pacific voyages, and the culture that produced the objects that he brought back with him as souvenirs to Leicester Square, London. Our bridge to that different knowledge and appreciation of sharks depends on these objects, as well as on the accounts of explorers such as Captain Cook.

References

Bernstein, P.L. (1996) *Against the Gods*. New York: John Wiley & Sons.

Boelhower, W. (2013) *New Orleans in the Atlantic World: Between Land and Sea*. London: Routledge.

Burke, E. (2009) *On Taste, On the Sublime and Beautiful: Reflections on the French Revolution. A Letter to a Noble Lord*. New York: Cosimo Classics.

Burn, G. and Hirst, D. (2001). *On the Way to Work*. London: Faber & Faber.

Cohen, M. (2010) *The Novel and the Sea*. Princeton and Oxford: Princeton University Press.

Ellis, R. (2012). *Shark: A Visual History*. Guilford, CT: Lyons Press.

Geary, C.M. (ed.) (2006) *From the South Seas: Oceanic Art in the Teel Collection*. Boston: MFA Publications.

Gessner, C. (1558). *Historia Animalium*.

Hurst, N. (2006) *Power and Prestige, the Arts of Island Melanesian and Polynesian Outliers*. Cambridge, MA: Hurst Gallery.
Kaeppler, A.L. (2008) *The Pacific Arts of Polynesia and Micronesia*. Oxford and New York: Oxford University Press.
Kamensky, J. (2016) *A Revolution in Color, the World of John Singleton Copley*. New York and London: W.W. Norton.
Kent, S. (1994) *Shark Infested Waters, the Saatchi Collection of British Artists in the 90's*. London: Zwemmer.
Kuzmanovic, N.N. (1995) 'The Shagreen Work of John Paul Cooper', *Antiques Magazine*. September.
Moyle, F. (2016) *The Extraordinary Life and Momentous Times of J.M.W. Turner*. New York: Penguin Press.
Neff, E.B. and Weber, K.H. (2013) *American Adversaries: West and Copley in the Transatlantic World*. Houston: Museum of Fine Arts.
Quammen, D. (2003) *Monster of God*. New York and London: WW Norton.
Raban, J. (1993) *The Oxford Book of the Sea*. Oxford: Oxford University Press.
Shaw, T.M. (1997) *What Manner of Men*. Ann Arbor, Michigan: University Press of America.
Wood, P.H. (2004) *Weathering the Storm*. Athens, Georgia: University of Georgia Press.
www.picturingamerica.neh.gov/downloads/pdfs/Resource_Guide_Chapters/PictAmer_Resource_Book_Chapter_10A.pdf

Websites

http://www.cartoonstock.com/sea0047
http://www.allposters.com/-sp/You-ve-got-a-bit-of-thong-caught-between-your-front-teeth-New-Yorker-Cartoon-Posters_i9168557
http://www.art.com/products/p47446786134-sa-i10619521/pat-byrnes-a-pet-always-boosts-a-president-s-approval-rating-cartoon.htm
www.arts.guardian.uk/saatchi/story/0,,928850,00.html
https://australianmuseum.net.au/blogpost/science/triple-beizam-hammerhead-shark-headdress-an-interview-with-ken-thaiday
www.davidshowalterart.com/section/394376_SHARK_PAINTINGS.html
https://condenaststore.com/featured/two-men-stand-on-a-desert-island-david-borchart.html
https://www.encyclopediavirginia.org/White_John_d_1593
http://www.everhartfineart.com/#/shark/
https://fineartamerica.com/products/port-jackson-shark-18th-century-natural-history-museum-london-art-print.html
www.jordanelgrably.com/blog/2015/6/11/photographer-graciela-iturbide-myth-and-matriarchy-in-mexico
https://media.newyorker.com/cartoons/593b61e78a032b339a0cd368/master/w_727,c_limit/110613_a15384_g2048.jpg
http://www.smithsonianmag.com/science-nature/sharks-made-out-of-golf-bags-a-look-at-the-big-fish-in-contemporary-art-23435794/
www.time.com/3927723/three-shark-photographers-you-need-to-follow/
www.abc.net.au/news/2016–02–01/vic-hislop-shark-show-closes-hervey-bay/7129802
www.nytimes/2006/10/01/design/01voge.html

www.teara.govt.nz/en/sharks-and-rays/page-2
www.perrotin.com/artists/Xavier_Veilhan2/8/the-shark/14409
https://i.pinimg.com/736x/05/ef/d2/05efd265d736d10f4c3c03c23a11d562—august—the-new-yorker.jpg
https://i.pinimg.com/originals/94/4f/1b/944f1b0d4e89adf868d6a9226b4e8296.jpg
https://i.pinimg.com/236x/37/af/5d/37af5db5843fdfe1cbe460b467e520a3.jpg
https://today.duke.edu/sites/default/files/styles/story_hero/public/sharks.jpg?itok=6h0i5smQ
https://static1.squarespace.com/static/535dab4ae4b055c269bff5ea/53656f96e4b0bfbb0aba7ddb/53656fd8e4b01e9923eb7aff/1399156698629/NYer_HumanitarianShark.jpg?format=500w
https://condenaststore.com/featured/two-men-stand-on-a-desert-island-david-borchart.html
http://indigenousjesus.blogspot.com/2011/11/artist-profile-safina-stewart.html
http://www.wildaid.org/news/sharks-humanity-opens-beijing

Other useful references

Abrams, A. U. (1979) Politics, Prints, and John Singleton Copley's Watson and the Shark. *The Art Bulletin.* Vol. 61, No. 2 (Jun), 265–276. https://www.jstor.org/stable/3049892?seq=1#page_scan_tab_contents

Allen, H. (1875) An Analysis of the Life-Form in Art. *Transactions of the American Philosophical Society.* Vol. 15, No. 2, 279–350. https://www.jstor.org/stable/1005378?seq=1#page_scan_tab_contents

MacDonald, J. (2015) Shark Attacks: A History of Hype. *JStor Daily*, (July 28). https://daily.jstor.org/shark-attacks-history-hype/

http://www.blouinartinfo.com/news/story/2268619/on-sharks-and-humanity-at-hong-kong-maritime-museum.

6 Prose sharks

In the eighteenth and nineteenth centuries, the shark became more than a name, more than a description in a book to the educated man and woman. Beyond the fisherman's quay and the coastal market, it was for the first time an object to view up close, even to touch. As early as the seventeenth century, shark specimens had been assembled and carefully arranged by John Tradescant and Elias Ashmole, but what began as a niche pursuit became part of mainstream urban culture over the next hundred years (Walton, 1784:25–6). Sharks began to appear among the furnishings in London coffee houses (e.g. *A catalogue of the rarities*, 1731:14). A shark's jaw was found among the auctioned collections of the late earl of Oxford in 1742 (Cock, 1742:8). A young shark was on display in the British Museum in the 1760s (Dodsley, 1761:95). Rackstrow's private museum in Fleet Street contained a young blue shark which 'with the assistance of Glass Eyes appears as alive' (*Rackstrow's Museum*, 1784:27). The Lewis Cabinet in Cambridge University Library, presented in the 1720s, contained the jaw of a shark, as did Richard Greene's collection at Lichfield in the 1770s. Clearly these things were sought out, notwithstanding Samuel Johnson's scepticism for the Cabinet of Curiosity: 'A man', he wrote, 'goes not away much illuminated by having enjoyed the privilege of handling the tooth of a shark, or the paw of a white bear' (*The Rambler*, 1752:iii, 109).

Sharks in a curiosity culture

The routine association of rapaciousness with the shark's jaws, found in Smollett and elsewhere at this time, might reflect the authors' familiarity with such sights as they went about their lives (Smollett, 1751:ii, 100). In both *Peregrine Pickle* and *Sir Launcelot Greaves,* Smollett suggests that cabinets of curiosity could topple over into props for charlatans, mere mumbo jumbo. In *Pickle*, an attempt to frighten Trunnion 'with an apparition' involved 'a leathern vizor of a most horrible appearance, stretched on the jaws of a shark' (Smollett, 1765:i, 103). In *Greaves*, the bogus 'conjurer' seeking to make a suitable impression on customers equipped his lair with, among other dead creatures: 'the jaws of a shark, and a stuffed weasel' (Smollett, 1762:ii, 202). Yet the ubiquity of the phenomenon tells us something. Sharks, for good or ill, impressed the observer with their latent power and danger. The lexicographer and scientist Benjamin Martin's improving dialogue between a young gentleman and lady, published in 1782, illustrates their impact on an impressionable audience. The girl asks the lad if he has ever seen a common shark. Not a live specimen, he replies, 'but I have seen the most notorious part of it, and that is, its tremendous Head and Jaws, which strike Horror into every one that views them, on account of their largeness, and the serrated Rows of Teeth in each, horrible to behold' (Martin, 1782:iii, 158).

Alongside a growing taste for the exotic and bizarre, and a growing awareness of the wider world stimulated by accounts of sea voyages into the Pacific and the Southern Ocean, the shark consolidates its place of dishonour among the exotic creatures and perils, entertaining a British reader by the fireside on a dark night (*Winter Evening*, 1753:ii, 273, 291). The defining characteristics of the oceanic shark remained its appetite, tyranny within its dominion, and opportunism. Familiar features and habits are rehearsed over and over in standard reference works: sharp teeth, appetite, a 'peculiar enmity' to man, size, the nurture of its young, movement in the muscles and heart after death, and a useful skin (e.g. Goldsmith, 1774:vi, 238–47). As the novelist and author of *Tom Jones* Henry Fielding noted, reference to sharks is by mid century 'exactly conformable to the rules and practice of voyage-writing' (Fielding, 1755:201). Indeed, the tall 'sea story' was already being spoofed effectively in the 1780s when a correspondent writing to the editor of *Wit's Magazine* ran through every cliche of the genre in the course of a few hundred words, including a concise account of being swallowed by a shark. This jolly narrative concluded with the usual disclaimer: 'If any one dares to dispute my veracity, let him seek ocular demonstration; for, be assured, no written evidence shall be allowed either to convict or confute me' (*Wit's Magazine*, 1785:ii, 16–17). Equally tall tales of live creatures existing happily in the shark's mouth or tummy were parodied in the anonymous 1790s work *Gulliver Revived*, where sailors catch a shark and open it, to find several partridges in the stomach, one of them sitting on a nest! (*Gulliver revived,* 1799:138–9). As another collection put it when printing the story, 'A traveller is allowed to lie' (*Jovial jester*, 1786:32).

Many traded on that permission. In a 'racy shark story' told by Senator John Spooner of Wisconsin in the 1880s, one shark is said to have swallowed 'a water pail, a child's rocking-horse, a pair of fisherman's boots, three empty beer bottles, a copy of The London Times, the well-preserved seat of a pair of trowsers, and a square piece of pine board, on which was painted the following sign: No Smoking Allowed' (*Milwaukee Sentinel*, 9 October 1887, p. 10). Some of these stories enjoyed a very long life. One with perhaps a ring of truth to it, about a shark, over nine feet long, caught in the River Thames in 1787 which was found to have a gold watch in its stomach, part of the effects of a young man lost overboard a couple of years earlier in the English Channel, was already old when reused in an article of 1822 ('Articles in the stomach of a shark', *The Kaleidoscope*, 22 January 1822: 231). This and the story of papers from the American brig *Nancy* recovered from the stomach of a shark, also dating from the late eighteenth century, were still cropping up in children's literature, and even in magazines targeted at a grown-up audience, a century later (for the *Nancy* see Thompson, 1970: 97–104). In the latter tale, the *Nancy*'s papers provided crucial evidence in a court case alleging that one ship had been masquerading as another to get around a trade embargo.

The implausibility of or exaggeration in these stories reflects the ubiquity of the shark in competing newspaper and journal literature. By the beginning of the 1870s, a periodical had to preface its story by asking: 'What on earth new can be said about shark-fishing?' (*Baily's Monthly Magazine of Sports and Pastimes*, 1 July 1872, p. 87). The novelty could, however, only take so many forms. Most attempts continued to focus on the shark's stomach. In 1840 a story went the rounds about a message in a bottle, found in a shark, signed by the Captain and crew of the missing ship the *Ontario* and reporting that their vessel was sinking and that they were taking to the lifeboats. The *New England Weekly Review* (Issue 34, 22 August 1840) was in no doubt that this was a concocted 'shark story' – note the tag – and that other papers had been taken in by a hoax; perhaps the identity of the 'special news-collector' who breaks the story, John Shark, was a bit of a giveaway? Something similar – another message in a bottle ostensibly from a ship in distress – was reported by the *New York Herald* in 1844 (Issue 238, 27 August 1844) as 'Humbug Probably'. In this case the message supposedly came from the 'brig, Shark'. Best of all, perhaps, was the story going about in the early 1870s in which a young man from the former Confederacy now enjoyed the delights of Paris, busily spending 'a fortune found in the belly of a shark, harpooned off the Carolina coast'. 'May treasure sharks be plenty,' the story concludes ('When my shark comes in', *Georgia Weekly Telegraph and Georgia Journal & Messenger*, 28 July 1874).

The trend culminated in some utterly far-fetched tales designed only to mock the genre. One was a story of a ship's carpenter and his son, swallowed by a shark and then resolving to cut their way free! ('The last shark story', *Cleave's Penny Gazette of Variety and Amusement*, 10 April 1841). Another parodied a gallant hero's improbable escape when sharks attacked: a tough, manly type, he blinds twenty sharks single handed ('A Shark Story: The late William P. Hawes Esq', *The Penny Satirist* 18 January 1845, p. 2). Another still concluded with the narrator eating the shark rather than the other way about! ('A 'tale' of a shark', *The Dart: The Midland Figaro*, 12 December 1884, p. 14). This absurdity was captured nicely in a delightful London cartoon showing a well-dressed shark perched on an urban rooftop, 'cleaning his boots and smoking cheroots', and singing of 'the ships he'd eaten like chips / In the palmy days of his youth'. Not at all bashful, this dandy 'carrolled away' so long, boasting of his romantic conquests among the whales and swordfish, that he made 'every one ill'. Like their publishers, some sharks never seem to know when to stop spinning a far-fetched yarn (L. E. R., 'The Shark', *St. Nicholas*, [1 March 1876], p. 312).

Another example of fiction wearing a very loose factual cloak, 'The Shark and the Landshark', neatly brought the human and the animal together in a tale of providential justice. 'Major Poppleton', the life and soul of a homebound East Indiaman, instant friend of everyone on board, has his leg bitten off by a shark in a freak accident. Shocked onlookers gradually realise that the shark has severed a 'cork leg', and congratulate the major on his escape. However, he is hardly consoled, declaring that he is ruined, and offering a generous reward for the return of his artificial limb. On the following day, the sailors capture and open the shark, and the leg, or what is left of it, is returned to its owner. It contains a secret compartment, and, despite the major's attempts at concealment, this is observed to contain jewels. Is it coincidence that there has recently been a spectacular jewel heist in India? In the end, the major profits from his apparent crime, but through a couple of twists in the plot, other more worthy causes benefit as well (*Young Folks Paper: Literary Olympic and Tournament*, 9 June 1888, p. 365).

Aware of the shark's reputation as the dustbin of the ocean, eighteenth- and nineteenth-century literature suggests that sailors and passengers took routine enjoyment in opening the stomachs of these fish, for who knew what might lie within? Amid the tedium of unpredictably long sea voyages under sail, the shark simply offered entertainment. Impressive dimensions were sure to prove interesting, such as the shark of sixteen feet recorded off Mauritius in the 1740s. Accounts abound of sharks, their meals, their habits and their size, reflecting the increasing number of deep ocean voyages written up and then repeated over and over, from official reports into published narrative, and then as extracts in papers and magazines.

'Of all the inhabitants of the deep, those of the shark kind are the fiercest and the most dangerous.' Thus began a 1778 article on the 'Natural History of the Shark'. This 'Natural History' exaggerated the size of the great white, and its appetite for man. In the burgeoning field of the popular literary magazine and the newspaper, that robust, partial view of the shark does not really evolve very much through the century and a half that follows. Shark literature is overwhelmingly designed to titillate new readerships created by wider educational opportunity and greater disposable income across the English-speaking world on both sides of the Atlantic.

In contrast to the variety of and ingenuity shown by the human shark, there was little subtle or novel in the many popular nineteenth-century treatments of shark species swimming in the ocean, whether published in periodicals, newspapers, or in some more permanent forms of literature. Sharks featured as monsters, worthy opponents for heroes, and they again and again cropped up as the pursuers, at times the dispatchers, of innocent swimmers. They offered a nautical sadist new forms of entertaining torture, such as the 'shark pit' described in the popular author of historical yarns, S. Walkey's serialised novel *The King of the Seas* (1899). Prisoners and other recalcitrants were lowered into this watery hole, where the sharks waited, working themselves into a frenzy of expectation (Walkey, 29 November 1899:234). All teeth, agility, power and menace, sharks both appealed to and appalled the reader. They provided a test of the imagination for a budding author's skills in a competition on a set theme, as in the case of a close encounter told by the fifteen-year-old John Peck in 1883, a prize-winner in the London magazine *St Nicholas* (Peck, 1883:710).

Tales of bloody encounters sustained the sailors' enduring belief that the shark was a creature of ill omen, and they continued to pursue sharks ruthlessly. In the naval officer John Rickman's account of Captain James Cook's last voyage, the taking of a shark 'left one tyrant the less to vex the ocean' (Rickman, 1781:18). The supernatural hold of the shark over the mariner is illustrated in an eerie tale of a British seaman serving on an American ship who foresaw his own death in the jaws of a shark, enacted below him one night as he looked on in horror from the ship's rigging. And indeed he died in this way before he saw his home again. The sailor's fate, the reporter concluded, 'added but another chapter to the psychological mysteries that happen from time to time, and which no man can explain' ('The ghost of a shark', 1890:6). Less spooky, but no less shocking, an account from 1829 recorded the deaths of two young men, when a boat lowered to watch the catching and dispatch of sharks gave way and tumbled twelve sailors into the water. One of the victims, in an affecting if somehow appropriate scene, slipped and flung his arms about the captured shark, already half raised out of the water. 'But when he found out what he had hold of, he was so horror-struck that he called [the author] by name, "Oh my God! My God! Witness my end!" and down he sank to rise no more' ('A shoal of sharks', 1830:119). It is possible that the shark fled with the man still attached, leading him to his death by drowning. Among many similar instances, a US artilleryman was reportedly eaten by a shark in Florida in 1857 (*Charleston Mercury*, 22 May 1857). And a sailor was lost while swimming to his ship off Fernando Po in Equatorial Africa in 1860 ('A seaman devoured by a shark', *Frank Leslie's Illustrated Newspaper* (New York), 8 December 1860, p. 36). There is at least one story of a British sailor avenging his friend's death at the jaws of a shark, off Barbados (told several times, e.g. *Jester's magazine*, November 1765:51–2). Courage matched against cruel nature always makes good reading, and it is startling how often the shark, supposedly so fearsome, comes off worse.

As in the centuries past, 'shark sport' in warmer latitudes frequently amounted to idle killing for want of something better to do. According to another author, citing one G. Bennett, 'The capture of one of these voracious animals frequently beguiles a tedious hour during a long voyage. Its struggles, when brought on deck, are very great, but a few severe blows on the nose soon disable it from further exertion' (W.G.C., 1835:108). Equally bored US Naval personnel blockading Manila Harbour in the War of 1898 had 'rare fun shooting sharks' ('Pastime on blockade duty', *The Butte Weekly Miner,* 2 June 1898, p. 15). It was this kind of representation of shark killing as 'fun' that helped to perpetuate the belief that no value attached to sharks, and that they had no other function than to provide sports fishermen with entertainment. Just the same, the Spanish War, reported on by a host of journalists with columns to fill, fuelled an American interest in sharks which prompted sporadic attempts to highlight their timidity, and even to be objective about their man-killing propensities, perhaps to reassure nervous relatives of the sailors back home (e.g. 'A plea for the shark', *Macon Telegraph,* 5 July 1898, p. 4, story taken from the *New York Times*).

In fiction, as in fact, shark peril sometimes brought out the best in people. After the appearance of Captain Frederick Marryat's *Masterman Ready* and R.M. Ballantyne's *The Coral Island*, a vast number of variants and spin-offs aimed at the young, such as Arthur Locker's 'Mat and Dick, or Life on a Coral Reef', serialised in *Little Folks*, contain what seems to be an obligatory chapter on the dangers posed by sharks (*Little Folks*, Issue 87, n.d.). Another example is Henry Frith's serialised *The Cruise of the Wasp* (*Every Boy's Annual* (London), [date unknown]; pg. 49; Issue LXXIV). There are many others. In the tropics, it seems, no young lad in a novel or yarn ever puts out in a boat without being visited by a shark. It is a rite of passage. The suspicion must be that jobbing authors assembling a sea story drew on the obvious tropes for fear of being criticized for lack of excitement by expectant editors, publishers and readers.

Sharks and increased acquaintance

The nineteenth century was the age of increased mobility for the masses. It saw advances in deep-water diving, and an enduring vogue for sea bathing. A holiday summer season was born, as factories closed for 'wakes weeks' and the inhabitants of American and British industrial towns and cities set off en masse for the coast. In 1880, building perhaps on its earlier prominence within some strands of the tall sea-story, the shark was being seen as a suitable substitute for the sea serpent in filling copy during the newspapers' 'dead-alive' summer, when other news was thin, Parliament and Congress were in recess, high society had retreated to its estates and the law courts were also on vacation. The 'Shark Scare', nine times out of ten an invented journalistic fantasy or an exaggeration, was noted more and more around this time, for example in the form of the 'Sayville Shark' in New York State during 1894 – another vigorous replacement to the tired old tales of the sea serpent – and during the warm summer of 1898 ('That Sayville shark', *Bangor Daily Whig and Courier*, Bangor, Me, 19 July 1894; *The Saturday Review*, 17 September 1898, 375–6). Seaside landladies and bathing machine operators execrated the beasts, and their concerns generated still more good copy. Most readers saw the fun for what it was.

For anyone who took the 'threat' seriously, periodic 'sharking crazes', such as that which swept Nantucket in 1886, more than balanced the score. Attributing sharking to women of fashion, 'Ladies', a paper reported '. . . fish daily in their bright holiday attire for them'. The resulting slaughter was considerable. In keeping with the spirit of the summer season, the same paper reported the railroad developer Jay Gould's alleged denial of a report 'that sharks in the North River nod to him as he passes in his yacht'. The branches of the species, it seems, readily recognised one another ('Shark Sense', *Rocky Mountain News* (Denver), 12 September 1886, p. 14).

Article after article insisted that there was money to be made from the shark, a great deal of money if some reports are taken at face value. Shark fishing generally seemed to be on the increase in nineteenth-century waters, commercially as well as for sport ('Shark fishery', *The Mirror of Literature, Amusement, and Instruction, Nov. 1822–June 1847,* 2(1), 16. Retrieved from http://search.proquest.com/docview/2826719?accountid=9851). Skin, bone, fins and flesh – all could be processed commercially. There was oil to be had from the livers, and, especially from the larger species, returns were profitable. The widow of M. Oliveros, who long enjoyed a near monopoly of the shark business, died in Paris in the 1890s, leaving $2 million. Oliveros, obituarists remembered, 'used to be fond of expatiating, at his famous dinner parties, on the usefulness of this fish' ('Sharks made him rich', *Denver Evening Post,* 10 May 1897, p. 2). Numbers sometimes make an inescapable point: human predation was already on a scale that, even in the days of pre-industrial fisheries, sets more lurid shark tales in perspective. According to the *Nantucket Mirror* in 1852, 'from eight hundred to one thousand' sharks, five to twelve feet in length, were taken in two months at Quidnit, Siasconset, and the south side of the island. 'Considerable quantities of oil are obtained from their livers, and the bodies are sold to the farmers of the island for manure' (reported in *Daily National Intelligencer*, Washington DC, 13 September 1852).

Opinions continued to differ as to the edibility of the shark. There were bon vivants who liked the fish, though species and size mattered ('Strange edibles', *Galveston Daily Press*, 24 June 1887, p. 8). Another writer distinguished between the 'blue' shark and the smaller 'brown' shark when it came to tastiness. The latter made a decent steak, according to sailors ('Captain Williamson's description of the shark' (1811), *The Weekly Entertainer: Or, Agreeable and Instructive Repository*, 6 January 1783–27 December 1819, pp. 51, 456. Retrieved from http://search.proquest.com/docview/4280556?accountid=9851). Sharks were themselves employed in a human sharking operation in 1890s San Francisco, when they were sold as 'tenderloin of sole' in the fish markets: 'carrion thus masquerading as respectable fish' ('Along the wharves', *Daily Evening Bulletin*, 27 September 1890, p. 4).

Often shown as dumb or unfathomable brutes, sharks were very occasionally credited with near 'human' intelligence, as in the stories circulating about the 'monster' fish San Jose Joe, a fixture off the Californian coast for many years ('A shark that thinks', *Aitchison Daily Champion*, 31 October 1889, p. 5). Some stories have sharks – albeit unconsciously – executing natural or divine justice. In one, a sorceress on the Sandwich Islands tells a husband eager to convince his wife of his fidelity and love that the sharks will arbitrate the matter. He must swim a certain course three times and, if innocent, the sharks will let him live. He survives, much to the surprise of both wife and sorceress ('Sharks as umpires', *Reynold's miscellany of romance, general literature, science, and art* 41.1062 (17 October 1868), 278).

As Britain's political and mercantile reach extended round the world, the periodical literature also ranged world-wide, full of shark escapes, shark killings and the occasional 'cool native' who, as in a case near Calcutta, traps the shark by coiling him in a rope ('Battle with a shark' (1823), *The Mirror of Literature, Amusement, and Instruction, Nov. 1822–June 1847*, 1(19), 300–1. Retrieved from *http://search.proquest.com/docview/2784043?accountid=9851*). Man has, of course, always challenged the shark's dominance, but from the early eighteenth century, as knowledge of other cultures developed, one begins to find reports of natives who can swim well, taking on the shark with a knife in its own element, and 'very commonly' killing him (Lawson, 1714:155). Pearl divers in Ceylon, armed with knives, were ready to fight back if attacked: 'a slash upward or downward has hacked the life out of many a fierce tiger of the sea'. Despite these killing skills, the Ceylonese divers felt safer still if there was a professional 'shark-charmer' in the boat ('Under sixty feet of ocean', *Chums*, [3 January 1900], p. 315). The villain in John Robert Hutchinson's novel for children, *The Quest of the Golden Pearl*, serialised in the 1890s in the magazine *Chums*, works as a 'charmer' of this kind.

Occasionally we find something different, such as an early example of the enduring story that sharks made use of the newly-opened Suez Canal 'for improper purposes', to hunt in new waters. Ten years later, patriotic US newspapers knew who to blame for this: French engineers. 'Thanks to M. de Lesseps,' readers were informed, 'shoals of sharks from the Red Sea and Indian Ocean glide through the Canal and infest the Mediterranean, where they are rapidly destroying the edible fish' (*North American and United States Gazette*, 2 September 1872; *Daily Evening Bulletin*, 26 August 1882).

An unwitting agent, again, and as we have seen in earlier chapters, the shark was enlisted in some elevated causes. From the later eighteenth century, popular outcry across Britain and parliamentary manoeuvres in London against the slave trade drew attention to the sharks which followed slaving vessels across the Atlantic, hoping to eat not only the waste from a ship, but also dead bodies discarded overboard (Grainger, 1766:76n; Rediker, 2008:285–97). The captain of a slave ship, in one possibly apocryphal story often repeated by the abolitionists, demonstrated to his human cargo that, despite the privations of their Atlantic crossing, suicide overboard was no trivial option – he assembled his cargo and dangled a female slave in the water, where she was bitten in two by a shark (Percival, 1777:25–6). Messages are seldom straightforward, however. Like other, similarly gruesome stories, this tale was repeated by both abolitionists and slavers for different purposes and different audiences.

Among more direct expressions of revulsion, this attention resulted in 'The petition of the sharks of Africa', a *jeu d'esprit*. Through their petition, sharks lobby the House of Lords, arguing with all necessary formality that abolition of the slave trade would seriously prejudice their interests. They express indignation that abolitionists, under the 'specious plea' of humanity, should seek to prevent the trade. Don't listen to the misguided House of Commons, they say. Don't suffer sharks to starve. The petition is an original use of shark stereotypes in the cause of satire and abolitionism; it seems to have circulated widely and to some effect.

Only once in a while does an article in a journal speak with cool reason, putting the shark 'menace', and even the fish's voracity, into perspective: 'Watch a man go through three or four courses of animal food, commencing with soup, and then compare him as a carnivorous beast, weight for weight, with the most voracious shark . . . Facts and figures will, I think, weigh all in favour of the fish' (F.G. Aflalo, 'Our sea fish', *The Boy's Own Paper*, 25 August 1894, p. 747). In much the same informative rather than sensationalist vein is an article in *Chambers' Journal of popular literature*, also from 1894, which plays down the shark's readiness to pursue humans. The author, familiar with many species of shark while in the

West Indies, knows of no authenticated case in which they have attacked humans. The stories told of them are, he insists, 'much exaggerated'. While voracious eaters, they are 'naturally cowardly animals', useful scavengers keeping the oceans 'clean' ('About sharks', *Chambers' Journal of Popular Literature, Science and Arts, Jan. 1854–Nov. 1897* (1 September 1894), 551–2).

By the 1890s one occasionally finds criticism of the sailors' 'savagely cruel' shark hunting, and an admission of the creature's utility, even in the serialised sea romance for children. Gordon Stables takes us in this direction in his *Frank Hardinge; or, from torrid zones to regions of perpetual snow*. An old hand among the crew tells lads at the centre of the story that, though ugly, sharks are 'useful in a way. They are the scavengers of our shores, they and the gulls.' And then of course they sometimes 'act as undertakers' (Gordon Stables, 'Frank Hardinge', *The Boy's Own Paper* 16 January 1897, p. 243). Towards the end of the century we find the first observations that sharks do not cope well with captivity, however large the tank. Even this story, however, dwells on the folly of the enclosed shark swimming forever round and round the tank in one direction only, pausing just to eat. The silliness of the brute creature takes precedence over some implicit sympathy for its predicament ('Sharks delicate in captivity', *Milwaukee Journal*, 12 February 1895, p. 3). Although we know now that sharks have complex and diverse brains across the species (Yopak, et al., 2010), earlier representations of sharks as stupid or silly, of course, perpetuated ignorant assumptions that helped to condemn them.

The novel shark

In his non-fiction narrative *Blue Meridian*, Peter Matthiessen describes the 'implacable, uncontactable shark' as historically denied 'the historicity, diversity and agency of wild animals'. That thought has across time informed a surprisingly wide range of fictional work on sharks. By the nineteenth century, sharks were simply regarded as they had been for centuries past: exotic, strange, death-dealing and 'tricky'. Monsters, predators, scavengers, sharks occasionally manifest that extreme expression of the 'other': they were gods worshipped by some inhabitants of the Pacific Islands and the West African coast. But to Western eyes, at least, the unfathomable otherness takes precedence over the godliness.

The children's literature of high Victorian England shows, however, that they are now also emblematic of empire, found alongside pirates, cannibals, faithful and faithless natives, and violent storms on the warm oceans all around Britain's possessions, flourishing in the tropical sea-lanes frequented by the world's greatest maritime power. We may see them as the guardians, and encouragers, of Imperial knowledge: of an awareness of the wonderful if challenging wider world, land and sea, that lay under British dominion. One article for children from the 1890s takes us on a sharks' tour of the Empire: New Zealand, Sydney, the Cape, Calcutta, Samoa, even the UK, recounting many narrow escapes and hair-raising tales (James A. Manson, 'Sea kings of the world', serialized in the periodical *Little Folks*, London, 1 January 1894).

In these narratives, sharks are passive entities, acting in accordance with assumed human motives, in keeping with the human view of wild creatures more generally in this period. Frederick Marryat's *Masterman Ready* (1841) has several shark encounters: Marryat's story relates the adventures of the Seagrave family, shipwrecked survivors on a Pacific island, assisted by a black servant and a veteran sailor, the eponymous Ready. It has sharks killing a pig and predating a cow, before the title character relates a grim tale of a soldier killed and eaten at St Helena, and the loss of a shipmate in Port Royal when fleeing a man-of-war: a standard moral tale of 'out of the frying pan and into the fire'. Didactic, like so many nineteenth-century Robinsonades, *Masterman Ready* culminates in the noble death of the sailor and the rescue of the family.

R.M. Ballantyne's 1858 novel *The Coral Island* has stood the test of time better, perhaps because it is better told. These are perhaps the first widely popular children's novels to place children as the sole heroes and principal protagonists. From the perspective of a fifteen-year-old, Ralph Rover, *The Coral Island* tells the story of three boys, the only survivors of shipwreck, marooned on a Pacific island. The boys do well enough, facing down the occasional natural or human adversary. As a glance at the periodical literature of the period shows well, *The Coral Island* inspired many imitators in the later nineteenth century. Like *Masterman Ready*, it echoes the seminal novel by Daniel Defoe, *Robinson Crusoe* (1719), and Johann David Wyss's entertaining, equally didactic, *The Swiss Family Robinson* (1812, published in English in 1816), but unlike *Crusoe*, which makes no mention of the creature, it includes a 'horrible encounter with a shark' early in the book, one of the first challenges faced by the boys, and in a form that, with variants, becomes almost obligatory for any seafaring yarn in the later nineteenth and twentieth centuries. The 'horrible encounter' comes when the boys are fishing from a log, which they are using as a very rudimentary boat. The shark, described simply as a 'monstrous fish', is inquisitive, and the boys are vulnerable for they are fishing, necessarily, with their legs dangling in the ocean. However, the shark is driven off by ramming a makeshift 'paddle' down its throat as it turns on its side to bite. In another passage, which again inspires almost obligatory variants in the years to come, Ralph is captured by pirates who threaten to 'introduce' him to the shark, by which they mean throw him overboard, although they never get around to doing so. Add to this *The Coral Island*'s encounters with cannibals, assistance from noble savages, and ultimate happy ending, and you have the Imperial sea adventure in a nutshell. Whereas *Robinson Crusoe*'s island lay off the mouth of the Orinoco, most Robinsonades were located, largely due to the opening of the Pacific to Europeans later in the eighteenth century, in that ocean. Novelists found advantage in placing the action in the most remote, exotic, and least known corner: they were free to take liberties in representing such spaces to their readers.

Moby-Dick (1851) is of course about a whale, and an obsession, but it features sharks too. Melville dwells on the sailors' hatred for the fish in his chapter 'The Shark Massacre'. In some unsettling pages he describes the consequent frenzy of sharks thrown back into the sea, feeding even on their own intestines, a sight that leaves the 'savage' Queequeg wondering at the kind of God that made such creatures (Leroux, 2004:56). It is important to note that this uncharacteristic shark behaviour is clearly driven by confusion caused by their ordeal at the hands of ruthless sailors, but the focus for Melville is on the humans. He describes a vengeful, superstitious terror that is recapitulated in subsequent narratives. Of particular pertinence is the night scene, where sharks tear into whales lashed to the ship's hull. This captures the terror those sailors must have felt sliding around atop an oily whale with huge predators feeding a few feet away. Melville had a whaler's low opinion of sharks, but not as low as his opinion of some members of our own species. In a delicious instance of counterpoint, the sharks feeding is synchronous with the execrable Third Mate Stubb, tearing into whale steaks below decks, and separated from the sharks only by the hull of the *Pequod.*

The Strife of the Sea (1903) by the American former mariner Thornton Jenkins Hains (1866–1953) is an interesting set of tales about sea creatures – including a whale, rats on a ship, an albacore and an albatross – one of which is titled 'Johnny Shark', and chronicles an imaginary life in the ocean (p. 251). Born in the South Atlantic, the 'hero' endures a perilous 'babyhood', surviving in an 'eternal war' between creatures of every kind (p. 257). His mother is killed by a swordfish. He learns by instinct to kill and predate, even killing whales. He feeds on offal at river mouths and grows big: 'full twenty feet in length' (p. 267). He mates, and chomps one of his offspring in half. There is an injection of human emotion at times: hubris brings the shark down. 'Now he felt a contempt for all other creatures' (p. 267). He eats a human diver and finds him tasty, but is then snared by a baited hook. His tail is cut off and he sinks 'slowly to the white coral below . . . then the scavengers of the reef came creeping up to do their work' (p. 273). It is a life-cycle story which tells of sharks' power and plight for what they are. Bert Bender refers to Hains' impeccable seafaring credentials. The *Strife of the Sea* is 'a unified collection . . . each of the human and animal characters acts out his own story in the larger biological plot of life' (Bender, 1988:110–11). It's a Darwinian book, a rare attempt to spin a fictional yarn from the fish's point of view.

The American novelist Morgan Robertson (1861–1915) is best known today for his story 'Futility'. Fourteen years before the sinking of the *Titanic*, this tells the tale of an 'unsinkable' passenger liner, the *Titan,* which hits an iceberg and goes down with most of those on board. In Robertson's less well-known collection of short stories *Down to the Sea* (1905), the tale titled 'The Shark' is interesting for its juxtaposition of an old-fashioned, superstitious mate intent on destroying a shark that follows the ship, with a young, career-oriented captain who considers his crew cruel and irrational in their hatred of the fish. Some among the crew are intensely superstitious, and they all grow more nervous by the day as ominous 'portents' mount up. Methods differ on ridding the ship of this unwanted company. While the Captain orders that the shark be fed until it is sated, the Mate surreptitiously poisons it by baiting pork with quicklime. Experience and tradition are pitted against inexperience and innovation. And the former – of course – wins.

Faroese short stories. Translated from the Faroese and the Danish, with introd. and notes, by Hedin Brønner, a rare collection of early Faroese stories, contains a short story by Sverri Patursson (1871–1960), 'Eliesar and the Basking Shark', first published in 1935, translated here by Hedin Bronner. Eliesar is 'cold as ice towards pretty eyes'. Handsome in youth, he is now nearing sixty, and knows the fishing grounds. He neglects his farm for his love, the sea. He is lucky on the water, and others are envious. A 'museum German' comes round, buying exhibits for his Berlin collection. One day Eliesar encounters a basking shark while fishing alone in one of the outer grounds. He is 'not entirely unafraid'. The shark follows him when he makes for shore, rubbing against his boat. He assumes that 'of course the shark would soon think of something better than merely to scratch herself against the sides of the boat'. But this remains supposition. He diverts the shark with his catch of cod, reviving himself from the weariness of rowing by swigging from his emergency brandy cask. Steadily, he lures the shark onto a sandbank and kills it, calling it an 'evil brute', a 'devil'. With the help of the storekeeper, acting as a middle man, Eliesar sells the corpse to the German for a handsome sum, and so turns the tables on his neighbours who believe that, for once, his fishing luck has deserted him that day. There are echoes here of Hemingway's Santiago. Patursson was an environmentalist and promoter of his native language – he translated *Robinson Crusoe* into Faroese – but his main motive for writing fiction was to make life in the Faroes better known, and comprehensible to less remote readerships. The shark's motives remain ambivalent, and almost irrelevant in the story. Strangely, this experienced fisherman does not seem to have encountered a basking shark before (Paturrson, 1972:35–45).

Ernest Hemingway's immensely popular last novel, *The Old Man and the Sea* (1952), tells the story of Santiago, an elderly Cuban fisherman who has been unable to catch anything. Short and focused, the book tells of Santiago's intense struggle, and developing relationship, with a large marlin, which he eventually kills and straps to the side of his boat. Imagining that his luck has turned, he dreams of the high price the marlin will fetch, but before he gets to port Santiago is beset by sharks, attracted by the carcass. Despite Santiago's desperate efforts, the relentless, voracious sharks keep coming, and in time strip the marlin back to a skeleton. Exhausted, Santiago struggles back to his shack on reaching port and falls asleep. Other fishermen are impressed by the size of the skeleton still tethered to his boat, and in a key moment Santiago's apprentice, Manolin, is reconciled to his master, whose luck – in one sense at least – has clearly turned.

Aside from the references to drinking shark liver oil and stabbing them in the head, Hemingway's description of sharks is startlingly original. He describes the large mako shark that attacks the marlin as having teeth that 'were shaped like a man's fingers when they are crisped like claws. They were nearly as long as the fingers of the old man and they had razor-sharp cutting edges on both sides' (Hemingway, p. 61). Although his descriptions are far from accurate they are extremely compelling in the way that they convey the overwhelming impression of this large shark to the tired old man as he tries to protect his catch.

The writing is taut and at times harrowing. Initially lauding *The Old Man and the Sea*, critics have been more divided since the author's death (Jobes, 1968). But on one point, most agree: sharks are yet again props in this story, inaccurately portrayed, bit-part players. While the book may perhaps be 'a critique of triumphalist hunting' which delivers only a skeleton as a trophy, there is only marginal sign of the elder Hemingway's shift towards 'greater ecological benevolence', at least in respect of the sharks (Hediger, 2008:36). In this novel they still represent elemental danger, and Santiago reacts against their predations ruthlessly. In contrast he expresses regret, respect, even 'love' for a 'brother' to his honoured opponent the marlin – albeit on human terms, and arising from his recognition of how little one may know of the life of a fish (Beegel, 2006:265). In the curious back-story of a failing old man, down on his luck, Santiago also loves the lions, seen on an African beach in his youth and remembered almost nightly in his dreams. For humankind, some wild creatures are, we are obliged to see, more sympathetic than others. Only when, in spite of himself, the old man dwells on sin, does he start to think of the first mako shark's nobility, and beauty, and lack of fear. Man, as the weary Santiago appreciates, amounts to little. And, after all, 'the sharks are all sharks, no better and no worse'. Unlike Hemingway's other spare literalism in a list of identities, one can only feel the truth of this statement (Beegel, 2006:239–40). If they play on memory in interesting ways, as Sergio Perosa argues, that is as playful as these creatures get in this novel (Perosa, 2010:31–6).

It is unsurprising, given the importance of the book in cementing Hemingway's contemporary status and subsequent reputation, that the secondary literature on *The Old Man and the Sea* is so extensive. One recent essay recalls a tremendous white shark called the Monster of Cojímar, captured in spring 1945, when a traumatized Hemingway returned to Cuba from the Second World War. In a letter about the 'Monster' he drew parallels between sharks and humans, suggesting that Cojímar's wartime shark industry and the role played by sharks as scavengers during America's war at sea made a profound impression on him. Hemingway's subsequent efforts to write about sharks, killing, and sin suggest that *The Old Man and the Sea* may be a story about the war in which 'the war is never mentioned' (Beegel, 2015:9–35).

The story of Peter Benchley's *Jaws* (1974) is in part the classic hero-monster fable, in part the story of an obsession (Adams, 2008:106, 111). The Long Island beach resort, Amity, is plagued by a killer white shark. After several fatalities, the town authorities reluctantly close the beaches in the peak holiday season and sanction an attempt made by a local shark hunter, Quint, to kill the creature. Quint, Matt Hooper, a marine biologist, and Martin Brody, the local police chief, set sail in Quint's boat, the *Orca*. This trio is beset by personal jealousies – Brody rightly suspects Hooper of having an affair with his wife, Ellen, and Quint and Hooper despise one another for their very different attitudes towards sharks. The great white destroys the *Orca*. Hooper is eaten, Quint drowns, but the shark dies of its wounds at the very moment of its victory, as it is about to devour the only human survivor, Brody.

Benchley's novel shows how sharks beset Amity by land as by sea: the predatory Mafia money that bankrolls the town mayor, the opportunist selling spurious tickets admitting gullible tourists to 'shark beach', even the predatory extramarital sex between Hooper and Ellen. Let us be under no illusions: the human land-shark remains as prolific and as inventive as ever. There is an attempt to see things from the shark's point of view; at first he seems an instinctive creature to the professional fisher Quint, without 'the brains of a dog', a 'dumb garbage bucket' (Benchley, 2012:269). But is he? Brody, watching the fish for the first time, thinks that the shark doesn't merely look mean, 'he meant to be mean. Like he knew what he was doing' (p. 287). Early in the final confrontation, Quint is obliged to concede the possibility of intelligence: 'Smart or not, I wouldn't know . . . But he's doing things I've never seen a fish do before' (p. 295). Still, we may wonder whether too much is built on coincidence and anthropomorphic tendencies?

Jaws is also effective in highlighting the powerlessness of officials confronted by knowledge of sharks off their coast. There are natural as well as jurisdictional boundaries to their authority. As Professor Jack Musick asked after the 2001 Virginia attacks, 'What the hell are you going to do?' Elected officials are always vulnerable since, as with other natural threats and disasters, the electorate hold them responsible, however irrationally, when 'things go wrong'. President Woodrow Wilson may have won a return to the White House in the 1916 election, but he lost the state of New Jersey (Eilperin, 2011:197–8).

Kiana Davenport's *The Shark Dialogues* (1994) explores the conflicts and challenges faced by Hawaiians living under American rule, through the lens of women in several generations of one family. The matriarch figure, Pono, identifies with and on occasion becomes a shark god, living for days at a time in the Pacific. She struggles to accommodate the increasingly multicultural world of her daughters and granddaughters, who marry out of the Hawaiian blood and introduce American and other cultural influences into Hawaiian society. 'Davenport uses the concept of hybridity to express the blended nature of the characters' lives and gives the concept a central place in the novel.' In the end, after many, many twists and turns, the older generation comes to accept the 'new' world of the hybrid (Spencer, 2005:21–8). Hybridity lends the novel a particular complexity, for Pono, while in one sense 'pure-blood' Hawaiian, is nevertheless herself the result of mixed unions, and 'Her claim to be native lies not only in blood but also in her power as kahuna [seer] and her supernatural relationship to the shark deities' (Toyosato, 2000:71–82).

In *The Beach* (1996), Alex Garland tells the story of a cosmopolitan beach community of foreigners living out what they plan as an idyllic existence on a remote Thai island. Told in the first person by Richard, one of the last to arrive on the island, the story charts the fragmentation of the group into factions, and an eventual collapse into grim violence. In one passage a shark attacks a three-man fishing detail, killing one member and gravely injuring another, a Swede named Sten. The third, Karl, is scarred by what he witnesses and becomes a morose outsider. Eventually, Richard and his friends decide to escape, and spike the food of the group, but things do not go according to plan. Local Thai cannabis farmers – with whom the group maintains an uneasy relationship based on the mutual need for secrecy – reveal to the community leaders that Richard has, unwittingly, long since given away the secret route to the island. There is a fight. Richard and his friends depart. A year on, Richard has returned to the UK and is ostensibly reintegrated into society, though the island has left its mark: 'I carry a lot of scars.' In some ways the fatal shark attack is a pivotal moment in the novel, where things begin to unravel. But in other ways the unravelling is already far advanced (Stephenson, 2005:369). The novel describes an unfamiliar impact of a shark attack, revealing the fault lines in what purports to be an ideal society.

South African novelist Henrietta Rose-Innes' *Shark's Egg* (2000), 'seeks to turn from the political spectacle in search of other domains of experience, the sunny surface of the beach is cast into shadow by the menacing threat embodied by the shark lurking in the bay' (Samuelson, 2015:133). It is a coming-of-age story linking two young Cape Town women, one the 'double' of the other, and their romantic lives. And it is also a novel which plays successfully with the 'spatial conventions of Gothic literature' and also the 'Gothic trope of the split self and of duality', in a sinister tale of developing possession and replacement against the brooding backcloth of a rambling house and a wild sea. When one woman finally escapes, towards the sea and the glittering day, are we looking at suicide? (Joseph-Vilain, 2012:68). At any rate, she appears to be embracing the dark unknown world in which the shark is only one element.

There is more of the dark unknown in Pat MacEnulty's *The Language of Sharks* (2004), a collection of short stories set in Florida, a key area for shark encounters, by an American novelist. The title story is about Joan, a graduate student whose life is drifting. Among many other escapades, her platonic friend Victor – a married man and a polar opposite – takes her out on a boat at night to catch stone crabs in the Gulf. They feed crabmeat to a shark. Having taken dope, Joan 'absently let her fingers scrape against the water and thought about sharks and crabs and the way the stars inhaled and exhaled the universe' (pp. 194–5). Joan and Victor 'always gravitated toward water' (p. 196). She recalls one night an incident of early domestic abuse in which her father pushes her face through a glass window, leaving her scarred. 'I've been sinking ever since,' she says (p. 197). A former boyfriend, Robert, is still interested, but cannot understand why she is so taken up with 'all this college crap' (p. 198). Victor gives her a shark-tooth necklace: '"Fits your personality",' he'd said, and she took it as a compliment' (p. 199). There is often something sexual in our imaginings and visualizations of sharks. In a disturbing denouement, Victor and his friend Ray take Joan away for a threesome, and she does not protest. He gives her dope: 'treats for the queen of sharks' (p. 200). 'She thought of a television show she'd seen recently where they made a nurse shark go docile by turning her over on her back' (p. 201). Joan realises that 'You reach a point where your needs become toxic.' She gives up her College course and flies off to Robert (p. 203). She realizes too that Victor had deliberately 'taken her to the bottom', and that there was only one way to go from there, as she herself had told him.

The old simplicity of treatment is left far behind in Steven Hall's *Raw Shark Texts*, published in 2007. Sharks here are highly complex, elusive creatures. The title plays on psychological Rorschach Tests, that is, inkblot recognition tests. Hall's debut novel is a work of meta-fiction which uses linguistic jokes and cultural references. The plot is hard to summarise, given its elusive, referential structure. It tells of an amnesiac, Eric Sanderson, rediscovering a past life through written clues he has left himself while evading a steampunk villain and the 'conceptual shark', or Ludovician, of the title, which 'feeds on human memories and the intrinsic sense of self'. This shark, yet again living up to the creature's voracious, relentless reputation, is still pursuing him. In a deliberate homage to *Jaws*, Eric and his colleagues – Scout, his lover and guide, and Trey Fidorous, a doctor who knew him in his past life – build a conceptual boat, set sail on a conceptual sea, and destroy the conceptual shark by throwing a laptop wired to a hostile megalomaniac network intelligence – Mycroft Ward – into its mouth. This spells death for Eric in real life, but he lives on with Scout in the conceptual world. *Raw Shark Texts* is full of visual conundrums and puns, which keep a dedicated online forum busy. Several pages, for example, constitute a flip-book sequence showing a shark attack. It might well be termed a 'slipstream' novel (Hayles, 2011:115–33). The writing 'hovers between futurism and nostalgia, between modes of mediation that at once link up with "new" electracies and look back at "old" strategies of printing, cutting, and assembling' (Wurth, 2011:138).

The sharks are still essentially functional agents in Canadian writer J. Kent Messum's bleak debut novel *Bait* (2013). This novel springs from the author's 'own fears of isolation and dependence' (from the author's website *jkmessum.com*, 2 March 2016). The reader encounters another case of memory loss. Six people wake up on an island in the Florida Keys, but cannot recall how they got there. All soon discover that they share a heroin addiction. Watched by four mysterious individuals on board a yacht, a note is found, and the plot turns on the impulse to swim to a cache of heroin on the neighbouring island, without being taken by the sharks that infest these waters. In some ways this is straightforward exploitation fare – it is hard to sympathise with the victims – but the motive behind a set-up keeps the reader guessing. The trouble here is identified by Robert J. Wiersema in the *National Post* review: even when the characters are broadly unlikable, you don't find yourself rooting for the sharks, just anticipating the 'seemingly inevitable carnage'. But while we cannot quite root for them, don't we as selfish, driven creatures all have more in common with the sharks than we might care to admit? (online review, 13 September 2013). Alternatively, as Romantics and Post-Romantics, aren't we always temperamentally on the side of 'spectacle'?

In his stream-of-consciousness conceptual novel *Shark* (2014), either a sequel or a prequel to *Umbrella* (2012), the British writer Will Self introduces us to Claude Evenrude, an ex-US airman who, in a nod to the *Jaws* movie, survived the *Indianapolis* sinking in 1945. Evenrude is an inmate in Concept House, an 'anti-psychiatric community' in Willesden. He is an unreliable witness – do we believe him when he claims to have flown on the *Enola Gay* to Hiroshima? – who gives a particularly shocking account of the destruction worked by the sharks as he watches from a life raft. Zachary Busner, the doctor in charge of Concept House, takes an LSD trip with inmates, and things predictably go wrong. The novel ends when he goes to the cinema with his son and sees *Jaws*, when the plotline of the film suddenly draws a rampant series of associations of reality and imagined reality into a moment of clarification.

Self's shark is again predatory, not in any way human except by association, 'an agent of pure desire and pure violence', as a review in *The Guardian* suggests. In this respect at least, it is a timeless construction. It is 'the opposite of irony' (Stuart Kelly, 28 August 2014).

Sympathetic treatments of sharks have been few and far between. What we can say is that, in prose fiction, the shark plays predominantly to a human fear of the invisible, powerful predator, the dark and unknown (Ritter et al., 2008:45–52). Sharks have fulfilled the role of the monster for centuries, and will continue to do so unless we take responsibility for the impact that representation has on the lives of 'others'.

References

Unless otherwise stated, primary resources are accessed from Gale Primary Sources: Eighteenth Century Collections Online, Gale Primary Sources: British Library Newspapers, Gale Primary Sources: Nineteenth Century UK Periodicals, Nineteenth Century US Newspapers, Readex: America's Historical Newspapers, ProQuest.

Adams, J. (2008) Plot taxonomies and intentionality. *Philosophy and Literature*. 32, 102–18.

Aitchison Daily Champion.

Anon. (1731) *A catalogue of the rarities to be seen at Don Saltero's coffee-house in Chelsea. To which is added, a compleat list of the donors thereof*, London.

Anon. (1786) *The jovial jester; or, Tim. Grin's delight*, London.

Anon. (1799) *Gulliver revived: or, the vice of lying properly exposed*, 8th ed, London.

Anon. (1803) Natural History of the Shark, from Dr. Goldsmith, and other eminent writers (1778). *Universal Magazine of Knowledge and Pleasure, June 1747–Dec. 1803*, 63(440), 231–3.

Baily's Monthly Magazine of Sports and Pastimes, London.

Bangor Daily Whig and Courier. Bangor, Maine.

The Bee.

Beegel, S.F. (2006) A guide to the marine life in Ernest Hemingway's *The Old Man and the Sea. Resources for American Literary Study.* 30, 236–315.

Beegel, S.F. (2015) The monster of Cojimar: a meditation on Hemingway, sharks and war. *Hemingway Review.* 34:2, 9–35.

Benchley, P. (2012) *Jaws*. London.

Bender, B. (1988) Sea-brothers: the tradition of American sea fiction from Moby Dick to the present.

The Boy's Own Paper, London.

Butte Weekly Miner.

C., W.G. (1835) 'The shark', *The Mirror of Literature, Amusement, and Instruction, Nov. 1822–June 1847*, 25(719). 308.

Chambers's journal of popular literature, science and arts.

Charleston Mercury.

Chums, London.

Cleave's Penny Gazette of Variety and Amusement, London.

Cock, C. (1742) *A catalogue of the collection of the Right Honourable Edward Earl of Oxford deceas'd . . .*, London.

Cooke, E. (1712) *A voyage to the South Sea, and round the world, perform'd in the years 1708, 1709, 1710, and 1711 . . .*, London.

The Dart: The Midland Figaro, Birmingham.

Daily Evening Bulletin, San Francisco.

Daily National Intelligencer, Washington.

Denver Evening Post.

Dodsley, R. (1761) *A general contents of the British Museum: With Remarks. Serving as a Directory In viewing that Noble Cabinet*, London.

Every Boy's Annual.

Every Week: a journal of entertaining literature.

Fielding, H. (1755) *The journal of a voyage to Lisbon, by the Late Henry Fielding, Esq*, London.

Frank Leslie's Illustrated Newspaper, New York.

Galveston Daily Press.

Georgia Weekly Telegraph and Georgia Journal & Messenger, Macon.

[Ghost] 'The ghost of a shark', *The Aitchison Champion*, 28 May 1890, p. 6.

Goldsmith, O. (1774) *An history of the earth, and animated nature . . .*, London.

Grainger, J. (1766) *The sugar-cane: a poem*, London.

Grose, F. (1788) *A classical dictionary of the vulgar tongue*, London.

The Guardian.

Hayles, N. K. (2011) Material entanglements: Steven Hall's The Raw Shark Texts as slipstream novel. *Science-Fiction Studies.* 38:1, 115–33.

Hediger, R. (2008) Hunting, fishing, and the cramp of ethics in Ernest Hemingway's *The old man and the sea*, *Green hills of Africa*, and *Under Kilimanjaro. Hemingway Review.* 27, 35–59.

The Jester's Magazine.

Jobes, K. T. (ed.) (1968) *Twentieth century interpretations of The Old Man and the Sea.*

johnsonsdictionaryonline.com (1755 ed.)

Joseph-Vilain, M. (2012) 'Something hungry and wild is still calling': post-apartheid Gothic. *Commonwealth.* 34, 61–70.

The Kaleidoscope.

Lawson, J. (1714) *The history of Carolina; containing the exact description and natural history of that country . . .*, London.

Leroux, J. (2004) Of sharks and pilot-fish: Melville's prophetic art and the dream of French America, a reprise and reappraisal. *Leviathan.* 6:2, 39–61.

Little Folks.

The London Reader: Of Literature, Science, Art and General Information.

Macon Telegraph.

Martin, B. (1782) *The young gentleman and lady's philosophy in a continued survey of the works of nature and art; by way of dialogue*, London.

Milwaukee Journal.

Milwaukee Sentinel.

National Post.

New England Weekly Review.

New York Herald.
North American and United States Gazette, Philadelphia.
Paturrson, S. (1972) 'Eliesar and the basking shark' in *Faroese short stories. Translated from the Faroese and the Danish, with introd. and notes, by Hedin Brønner*, New York, 35–45.
Peck, J. (1883) 'A shark in sight', *St Nicholas* (London), [1 July 1883], p. 710.
The Penny Satirist.
Percival, T. (1777) *A father's instructions to his children consisting of tales, fables, and reflections; designed to promote the love of virtue, a taste for knowledge, and an early acquaintance with the works of nature,* London.
Perosa, S. (2010) Memory and the sharks, in M. Cirino and M. P. Ott (eds), *Ernest Hemingway and the geography of memory*, Kent State UP, 31–6.
[Rackstrow's Museum] *A descriptive catalogue (giving a full explanation) of Rackstrow's Museum: consisting of a large and very valuable collection of most curious anatomical figures, and real preparations: also figures resembling life*, London, 1784.
The Rambler, volume 3 (1752).
Rediker, M. (2008) 'History from below the water line: sharks and the Atlantic slave trade', *Atlantic Studies* 5, 285–97.
Reynold's miscellany of romance, general literature, science, and art.
Rickman, J. (1781) *Journal of Captain Cook's last voyage to the Pacific Ocean, on Discovery; performed in the years 1776, 1777, 1778, 1779*, London.
Ritter et al. (2008) E. Ritter, K. Lutz and M. Levine, When humans and sharks meet, in *New developments in the psychology of motivation.* ed. F. M. Olsson, New York, 45–52.
Rocky Mountain News, Denver.
St Nicholas, London.
Samuelson, M. (2015) 'Literary inscriptions on the South African beach: ambiguous settings, ambivalent textualities', in U. Kluwick and V. Richter, *The beach in Anglophone literatures and cultures: reading littoral space. 121–38.*
The Saturday Review.
The works of Shakespear in nine volumes . . ., London, 1747, ix Glossary.
'A shoal of sharks', *The Kaleidoscope: Or, Literary and Scientific Mirror* (1830), 11(537), 119. Retrieved from http://search.proquest.com/docview/2769264?accountid=9851.
Smollett, T. (1751) *The adventures of Peregrine Pickle. In which are included, Memoirs of a lady of quality. By the author of Roderick Random*, Dublin.
Smollett, T. (1765) *The adventures of Peregrine Pickle. In which are included, Memoirs of a lady of quality,* London.
Smollett, T. (1762) *The adventures of Sir Launcelot Greaves*, London.
Spencer, S. (2005) Cultural hybridity in Kiana Davenport's Shark Dialogs. *Storytelling.* 5, 21–8.
Stephenson, W. (2005) Island of the Assassins: cannabis, spectacle and terror in Alex Garland's *The Beach. Critique.* 46, 369–81.
Thompson, E. K. (1970) Tale of the "Nancy" brig. *Mariner's Mirror.* 56, 97–104.
Toyosato, M. (2000) Land and Hawaiian identity: literary activism in Kiana Davenport's *Shark Dialogues. Bucknell Review.* 44, 71–82.
Walkey, S. (1899) The king of the seas. *Chums.* London, England.
Walton, I. (1784) *The complete angler, or contemplative man's recreation; being a discourse on rivers, fish-ponds, fish, and fishing*, 4th edn, London.
The Weekly Entertainer: Or, Agreeable and Instructive Repository.

The winter evening's companion: or, compendious library. Being a collection of instructive and entertaining relations . . ., London, 1753.

The Wit's magazine; or, Library of Momus. Being a compleat repository of mirth, humour, and entertainment, London, 1785.

Wurth, K. B. (2011) Posthumanities and post-textualities: reading *The Raw Shark Texts* and *Woman's World. Comparative Literature* 63:2, 119–41.

Yopak, K. E., Lisney, T. J., Darlington, R., Collin, S. P., Montgomery, J. and Finlay, B. (2010) Brain scaling from sharks to primates: a highly conserved vertebrate pattern. *PNAS*. 107, 12946–12951.

Young Folks' Paper, London.

7 Sharks on film

Shark movies

Shark movies are among the most consistently lucrative at the box office, and the reason is simple: sharks are among the most visually compelling and naturally spectacular of all animals. Unfortunately, success in the film industry tends to breed imitation, including the repetition of tropes, which can then become clichés. Some of the tropes of shark movies include a large black shark fin glimpsed in an early scene, foreshadowing attacks to follow; the jump scare, with the shark suddenly and terrifyingly appearing; the myth of the persistent predator, the shark that develops a taste for human flesh; the nerdy scientist who proves himself or herself to be heroic as well as brilliant; and the karmic death, in which a villainous character meets his fate in the jaws of a shark. All of these tropes are familiar to students of horror films, which is no accident since shark movies so often cross over into that genre. In this chapter we deal with some of the more successful and influential shark movies and dissect some of these representations to demonstrate how accurate or distorting they are in depicting the true nature of the top predatory species of sharks. We consider some of the many ways in which movies simplify sharks, thereby deliberately distorting public opinion about them, through a variety of visual tropes and schemes, some of which are common to thriller horror movies and others that are specific to shark movies (Figure 7.1).

We should begin with a very brief account of the general structure of movies, which usually begins with a screenplay, written in three acts, and covering a period of around 120 minutes. Each act has a different role to play in the movie. Act I is the establishing act in which environment, protagonist, premise and plot are all set up. In this act we see the protagonist typically in his or her comfort zone to give the audience a sense of how things usually work. The establishing act typically runs for about 30 minutes, with a plot point occurring just before that time limit is reached, to take the protagonist out of his or her comfort zone. This event will precipitate Act II, which usually lasts around 60 minutes, broken into two halves with a mid-point event that is often a false climax, a moment of relief or success of some kind before the protagonist is confronted with fresh challenges. Throughout Act II we see the protagonist on a journey that shows us what he or she is really capable of. Immediately before the end of this act, something has to happen through a plot point that will turn events around so that Act III can do its job of bringing the protagonist's journey, internal and external, to an end, through a series of scenes linked by a common theme. Of course, there may be many plot points throughout the screenplay, but the Act I plot point, the mid-point and the Act II plot point are prescribed as the crucial turning points on the protagonist's journey.

We can take Steven Spielberg's *Jaws* (1975), still the most successful shark movie, as an example. It includes all of the tropes because it invented many of them (and did so wittily). *Jaws* adheres clearly to the three-act structure of set-up, conflict and resolution, even while it has been variously dissected as having a three- and four-act structure (Connor, 2015). In Act I of the screenplay, after a midnight bather is discovered in pieces on Amity's beach, Hooper (Richard Dreyfuss), a shark expert, is called in to investigate. At the same time, the people of Amity call for the shark slayer Quint (Robert Shaw) to track and kill the shark for a large sum of money, while Chief Brody (Roy Scheider), the aquaphobic police chief of Amity Island, attempts to do the right thing by keeping himself and everyone else as far away from the water as possible. Act II of the movie puts this incongruous threesome on a boat – an even smaller island – in pursuit of the shark. The relationship between the hunter and the scientist is charged with conflict. Brody, whose only aim is to return peace to Amity, is forced to confront and overcome his fear of water, sharks and Quint. Act III of the movie shows us how Brody becomes the hero by killing the shark. Although the struggles that Spielberg had with his mechanical shark, which he named Bruce, are well documented, the fact is that audiences are thrilled by it. The lasting images of the 'life-like' shark in the movie show it being blown to bits while Brody and Hooper rejoice. Only with the complete destruction of the shark can they swim back to land with the good news that Amity is once again safe for tourists.

Jaws depends upon a range of strategies for generating tension in the audience, including the famous John Williams score. As Emilio Audissino points out, the shark really doesn't show up much: 'In the film's 119-minute running time, the shark can be just glimpsed at 60 minutes and finally shows up only at 78 minutes. A crucial doubt now haunted the director and producers: in the few shots in which it appeared, would the fiberglass-and-rubber puppet really be credible and therefore perceived as a menacing shark?' (Audissino, 2014:110). Audissino goes on to explain how the music of John Williams, which begins with an excited heartbeat tune, is what made the imagined shark, a puppet, not merely believable but scary. In the opening sequence the music accompanies a shark point-of-view scene, causing the audience to think it has seen a shark in pursuit. This is a thriller-movie trope that is frequently used in shark movies to quickly establish the presence of a predator. As Linda Maria Koldau explains, it is much more complex a process than it seems: 'music (and sound) only unfold their true effects if complemented by the other side of a shark's attack – by sudden silence and the horrifying tension building up from it, in expectation of the next attack. Only then do music and silence add up to a convincing acoustic representation of this stealthy predator.' Matching verbal, visual and auditory emotional signals is highly complex. If any of these elements are out of synchronisation the effect is ruined.

After the opening POV sequence, the action begins with a girl, Chrissy, stripping off her clothes and going for a night swim out toward a lonely ocean buoy. When she is attacked we see only a body being manoeuvred around the water in an unnatural way. She screams, but we still don't glimpse a shark, even as she clings in terror to the buoy and then is yanked below the surface of the black sea. We will not see Chrissy again until she is discovered in bits on the beach, horrifically being picked at by crabs. The opening sequence demonstrates a skilful manipulation of a commonly experienced fear of the unseen, through a well-paced interaction of auditory and visual strategies.

Brody, the protagonist of *Jaws*, is used to cue the audience to fear sharks throughout the screenplay. His fear grows stronger as he witnesses a boy, a dog, and later his own son almost seized by the shark. His efforts to close the beaches are blocked by the self-serving and powerful town mayor, in an increasingly unflattering view of officialdom that is reflective of growing discontent in the American political arena more generally (Wasser, p. 71).

Jaws treats the appearing and disappearing shark-fin trope masterfully to generate fear as well as humour about fear. Having seen the horrifying effects of the shark, we expect the worst when we see the fin in the water off a crowded beach. People panic, lifeguards react with great alarm, guns are trained on the fin, ready to shoot to kill, till a young boy bobs up exposing his shark-fin hat prank. Our suspense dissipates in a fizzle of dramatic disappointment and relief.

Brody is delighted when a shark is captured, until Hooper tells him it is *a* shark, but not *the* shark – thereby rebuilding tension. Hooper drags Brody onto a boat at night, again for maximum scare effect, to explore the shark's territory. They come across a capsized fishing boat either staved in below the waterline or with a bite taken out of it. As Hooper investigates, a blur of white bobs into view. We lurch, but the white blob turns out to be not the expected shark but the body of Ben, a local fisherman whom the shark has surprisingly not eaten. It has, though, left a calling-card tooth to give Hooper a sense of its enormous size. This moment is doubly clever since it fulfils the trope in an unexpected way and references an actual instance of a boat being staved in by a white shark, off Nova Scotia, in 1953. Seeing Hooper spooked by Ben's body only contributes to Brody's growing terror. By the time Brody climbs aboard the boat with Hooper and Quint in Act II we already know that he is the least equipped to survive. This writing of the protagonist as a 'loser' is crucially important for the character arc that will see Brody transformed into a hero by the end of Act III. All of this character and plot-led emotional dynamic is written to ensure that the audience is on this journey with Brody and shares his fear of the monster, so that they will rejoice at the final image of Brody fearlessly swimming in the open ocean back to Amity.

Jaws was so powerfully written that people actually thought they had a licence to eradicate 'the problem' of the great white shark. 'The film has been blamed for the surge in shark game fishing, shark tournaments in the U.S. and worldwide overfishing that has devastated shark populations' (Francis, 2012: 57). The world population of great whites became endangered, to the point that governments all over the world had to step in and ban hunting operations. Western Australia refused to allow white shark tourism, though this decision didn't preclude a highly unpopular culling programme, allegedly to protect swimmers following a spate of unfortunate incidents at several of its most popular beaches. There are yet many Western Australian advocates for shark tourism as one important way of changing negative attitudes (Catlin, Hughes et al., 2014).

The Reef (2010), written and directed by Andrew Traucki, is very loosely based on an incident near Townsville in 1983, when a prawn trawler, *New Venture,* capsized with three people aboard. After 24 hours in rough sea conditions, two of the three were taken by a juvenile tiger shark, while the third kicked it away and was rescued after having spent 36 hours in open water (Healy, 2011). Traucki had already made a movie based on a true story of a deadly encounter with a crocodile, *Black Water* (2007), when he seized on this story for *The Reef.*

Act I of *The Reef* begins with Luke, the protagonist, meeting his friend, his friend's girlfriend, Suzi, and his sister, Kate, at the airport. He tells them that he has to deliver a boat to Indonesia and rather than hold him up they agree to go with him. This is the inciting incident that means they are all going to be in one space for the rest of the movie. The rest of this act sets up the relationships between two couples, with Luke and Kate growing ever closer. There are numerous shots of unhampered snorkelling and swimming in clear blue water, before the turning point at the end of Act I when the keel of the boat is broken on the reef and the boat capsizes. Their fortunes turn with the boat. Act II is where the story really begins as they each try to cope with being vulnerable in the middle of the ocean. Luke dons his mask and dives into the submerged cabin to retrieve what he can, as the others watch for sharks. At one point they bring him to the surface to warn him that there is something in the water. He tells them not to bother him again unless it is a big shark. He returns to the water, until now confident but increasingly nervous. He hears something banging against the cabin door. His hand reaches through the door, inviting the audience to believe that it will be bitten off; instead he pulls out the EPIRB (Emergency Position Indicating Radio Beacon), which has been knocking. This is a typical strategy for creating suspense and relief. The second major plot point occurs when Luke suggests that they abandon what is left of the boat and swim ashore. His co-pilot, Warren, refuses to take the risk, warning the others that in their wetsuits they are going to look like seals, sharks' favourite food. Nevertheless, the two couples leap into the water, each with half of a bodyboard as a flotation device. This may seem like a nice indicator of how sharing and caring they all are, but it is also a foreshadowing of their fates once they begin to splash on the surface of the water. The turning point happens at the end of this scene when Luke and Kate find a bit of rock within spitting distance, it seems, of Turtle Island. Her foot is bleeding, and as Luke binds it they declare their love for each other. From this moment we anticipate that Luke will risk his life to save her.

Act III begins when they leap from their love rock to swim the final stretch to the island. As the shark approaches, Luke pushes Kate onto the island and begs her to pull him up, but the rock is slippery and the shark snatches him away, leaving her to scream after him on her rock. Attempting to cling onto something useless for one's salvation is yet another thriller trope that is used several times in this movie, from slippery boats, to slippery rocks and here most effectively of all in Luke's reliance on Kate's slippery footing. An epilogue tells us that she was later rescued.

The plot is fairly well paced but there is very little swimming on this quest to find the island. The four huddle together in open water, the girls either shivering or panicking, while the guys offer false reassurances. Matt is the first to get his legs bitten off, heroically trying to retrieve Kate's lost flotation device and, again, heroically tells the others to leave him because the shark will be coming back. They refuse to leave him, but it turns out he was right – it does. What is effective throughout the movie is Luke's savviness. He pops on his diving mask and ducks underwater so that the oncoming shark sees that he has seen it. The shark then swims away for a while. As a wild animal that has to hunt very swift prey – primarily seals and dolphins – it relies on stealth and surprise. What is annoying is that Luke could have escaped attack if he had done in his final moment what he had been doing all the way through the movie, ducked and swerved to give himself time to get onto the island – but that would not have made a good dramatic climax. More satisfying would have been if he had scrambled up the rock, causing Kate to fall in head first – but then his identity as the heroic protagonist would have been destroyed. Screenwriters consider all of these options before they choose what works best for the movie's aims. Thankfully, the director avoided sensationalised shots of the shark's teeth with flesh and clothing dangling, avoiding the cliché.

The Shallows (2016), written by Anthony Jaswinski and directed by Jaume Collet-Serra, is the shark movie version of a recent spate of movies that focus on the lone warrior, hero and saviour, usually an A-list movie star and quite attractive, in a hostile environment. *All is Lost* (2013) features Robert Redford trying to cope with rough seas in a dinghy; *Gravity* (2013) features Sandra Bullock lost in space when her shuttle is destroyed; *The Martian* (2015) features Matt Damon who is falsely presumed dead by his crew who then take off, leaving him to manage life on Mars. The more recent *Passengers* (2016) offers a double lone-warrior scenario featuring Chris Pratt, who finds himself alone in a space station when he is accidentally awoken from hibernation, and Jennifer Lawrence, whom Pratt selfishly awakens to share his solitude. In *The Shallows*, Blake Lively is stranded on a rock, within sight of shore, while a great white shark circles her almost to dizziness. The juxtaposition of beauty and the beast is a common enough trope of horror movies, which this one also fulfils, as amply as Blake Lively fills her bikini. This film also satisfies another generally compulsory trope of shark movies, the outrageous distortion of shark behaviour.

The premise of *The Shallows* is a stretch: a white shark haunts the shallow water off an isolated Mexican beach, gobbling one male surfer within seconds of the last. The sole survivor is the female protagonist, Nancy, who proves to be more resourceful and more than a match for the implausibly energetic shark. Immediately after consuming not one but two sizeable male surfers, the shark swims at top speed back to Nancy. As she fends off this unbelievably persistent predator, she climbs onto a rock, where she bandages her leg with the shreds of her wetsuit. A blood-spattered bird teeters on the edge of her rock to add an extra symbolic and dramatic dimension to her predicament. As the tide rises around her rock, Nancy decides to swim to a buoy – a shout-out to *Jaws* – where she clings and wobbles until the shark monster dismantles it. In the final moments of the movie she cleverly lures the shark down to the spiked anchor point on the bed of the ocean, where she ducks just in time to see the shark monster impaled. The character fulfils the horror movie trope of the indomitable female Odysseus, recalling the role models of previous horror thrillers like John Carpenter's *Halloween* and the *Alien* movies. Our hero in *The Shallows* shows ingenuity and courage, and following a protracted ordeal, she, of course, prevails over the shark, even as the camera panders to male voyeurism (Mulvey, 1989; 2009).

Breaking down the film structurally, we see that Act I establishes Nancy as a young medical student whose mother has lost her battle with cancer. Disoriented, Nancy drops out of school and goes surfing instead. After hours of pleasurable uninterrupted surfing, the Act I turning point comes when a great white shark rams her surfboard and she is thrown into the ocean. Act II begins with her speedily and dramatically dragged by her surfboard, banging and slicing her leg on the rocks on the way. She scrambles to the closest thing to hand, a nearby whale carcass – a favourite food of great whites, which she now reeks of. She screams to shore at two surfers who are just setting off in their jeep. The jeep stops, raising hope of rescue, but the boys are just securing gear in the back – they haven't heard her. They drive off and the shark lunges at the dead whale. She sees a rock and leaps into the sea, only to step on coral spikes as she hoists herself out of danger.

However clumsy she is, Nancy is soon to demonstrate her formidable survival skills. She uses what she has learned at med school to staple her earrings to her gaping leg wound and then ties her sharkie (neoprene T shirt) around her leg for compression. This later appears as a neatly torn off sharkie arm that fits perfectly over her slender leg. After numerous shots of Nancy with a photogenic wounded seagull, the mid-point or false climax arrives with a homeless man wandering onto the beach. Having ignored her pleas for help, he steals her gear and then greedily wades into the sea to take her board. Almost immediately we see the man scrambling to shore with his legs detached from his waist – satisfying both the karmic death trope and the audience's desire for visual gore. When the two surfers seen earlier return, Nancy also alerts them, but as guys who know better they assure her there are no sharks here. The boys lie on their surfboards paddling, and in a POV shot we glimpse them looking seal-like from beneath the surface. Immediately the great white breaches and seizes one of the boys, employing the familiar Air Jaws attack strategy. The other boy starts paddling frantically and then bobs beneath the waves. His broken surfboard flies into the air. The boy scrambles to the rock, but the speedy shark snatches him away and the ocean turns deep red as Nancy looks on.

As she dries on her rock in the baking sun, long shots show how temptingly close she seems to shore. She uses the broken surfboard as a sun-shade. A GoPro drifts to her rock and she replays the footage of the attack. She freeze-frames the shark's jaws on the GoPro and sees a hook jammed in its flesh, registering that this shark seeks revenge. Having set her seagull adrift with a GoPro SOS message, she slips quietly off her submerging rock and swims for the buoy through a forest of jellyfish tentacles. She reaches the buoy, but the metal grab comes off in her hand. Desperate hand to jaws combat ensues, but Act II ends with a glimmer of hope as a small boy is shown finding her GoPro on the beach.

Act III begins with her shooting a flare at the whale oil, which immediately and unrealistically ignites, setting fire to the shark in another shout-out to *Jaws*. But the shark dives undeterred and attacks the buoy with even more determination, hurtling to the surface and ripping the buoy to pieces. She heads down to the ocean floor, faster than the shark by inches. She ducks away and the shark is impaled on the spikes devised for its demise. The boy returns with his father, who rushes into the sea to recover the unconscious body of Nancy. She spews water and checks she still has her body. The seagull stretches its wings on the beach while she hallucinates her smiling mother. One year later we are shown a close-up of a scarred leg on a beach. The final moments of the movie show Nancy and her sister heading out to surf, unvanquished and apparently untraumatized. Film critic Tim Grierson reads Nancy's confrontation with the shark as symbolic of her confrontation with her mother's death (Grierson, 2016). However, there isn't enough writing in place to really sustain that idea. Nancy emerges, quite simply, stronger having had to fight for a life that she had given up on.

When the movie was released as a DVD, it included an antidote to the misrepresentation of the monster shark by way of an interview with a naturalist moviemaker. However, as we already know from studies of Shark Week videos incorporating celebrity conservation announcements, the powerful fear instilled by horrific videos overwhelms the effectiveness of any attempts to offer the facts about shark attack (Myrick and Evans, 2014). It has been argued 'that familiar film narratives can serve as the basis for political discourse when they appear to mirror well-known stories, blame marginalised target populations and provide quick political "solutions"' (Neff, 2015:115). Movies are a much more common point of cultural reference than research papers by informed scientists. However, filmic representations are already imbued with writers' and directors' agendas to please the masses, but within the accepted context of a fictional illusion – a two-hour magic act, a sleight of hand which often includes several levels of computer graphic cutting and pasting of fiction onto reality to make it sell.

Humorous horror shark movies

Some of the shark movies clearly influenced by *Jaws* choose to undermine rather than imitate the strategies used to provoke fear in the iconic movie. The humour in horror movies is, as Philip Brophy points out, 'mostly perverse and/or tasteless, so much so that often the humour might be horrific while the horror might be humorous' (Brophy, 1986:12).

Deep Blue Sea (1999), written by Duncan Kennedy and Donna and Wayne Powers, and directed by Renny Harlin, opens with a group of four teenagers making love on a catamaran on a calm ocean at night. A bottle of wine is spilled, then they hear something pounding the boat. In one of many comic foreshadowings, we see a teddy bear sitting on top of the stereo fall into the ocean. A giant mako shark then bursts through the middle of the catamaran and the teenagers are thrown into the ocean. After much screaming and effort to scramble back up to the hull, a couple of harpoons are fired at the shark from a conveniently placed yacht. We later learn that this shark is an escapee from an underground scientific facility that has genetically engineered mako sharks to have bigger brains so that scientists can tap into their protein complex, which they believe contains a cure for Alzheimer's. These sharks don't just have bigger brains, they have bigger everything, running to 45 feet in length. They soon outwit the scientists, who fall prey to them in quick succession. The sharks breach the facility, causing flooding which in turn enables them to move freely along corridors and into offices to snack on geneticists. Only two scientists survive, by blowing up the one remaining supershark in an intentionally kitschy recapitulation of the flying blood and guts of *Jaws*'s finest non-conservational moment.

The best of the humorous horror shark movies, however, is the outrageously unrealistic *Sharknado* series (2013–17) written by Thunder Levin and directed by Anthony C. Ferrante. The conceit of these films includes sharks flying out of the sea on a vengeful mission that is clearly designed to elicit laughter and realise the parodic potentiality of *Jaws,* without trying to be *Jaws*. A monstrous storm afflicts Los Angeles, flooding the streets and even the interiors of houses with ocean water thick with sharks. The storm is so extraordinary, in fact, that hundreds of sharks are swept up into waterspouts, producing snapping, flying sharks whirling through the air. *Sharknado* features all the tropes, echoes and clichés it can muster, both from shark movies and B-grade disaster movies, beginning with the bosomy blondes and extending to bloody gore and chainsaws. Pushing believability over the edge is a simple strategy for creating laughter, and for establishing meaning by identifying where real silliness reigns, in this case the shark movies that take themselves seriously. *Sharknado* draws attention, usefully, to the absurdity of the monsters that we create for ourselves out of the shy creatures of the ocean.

Sharknado's success spawned a franchise, including four more movies to date. In the newest version, *Sharknado 5: Global Swarming,* the scene of a protagonist wrestling an open-jawed great white shark on the staircase of Buckingham Palace offers us amusing visual incongruities, while suggesting political realities, bringing together the land and oceanic shark visually, in keeping with a long history of literary representation. It also reveals that viewers are intrigued, as well as thrilled, by intrusions of the outrageously, spectacularly inappropriate into familiar spaces, particularly domestic spaces like kitchens and staircases. Beyond the humour of incongruities lies the spectacular and surreal experience, made safe by cinema and even safer by humour, of viewing the monster brought home. It's harder to be solemnly afraid of the monstrous image of the great white shark once you've seen it thrashing in the flooded dinette.

Bit parts for sharks

Hundreds of movies have incorporated sharks to serve their agendas in various ways, such as adding a ready-made thrill or picking up the action during a lull. Although the sharks are incidental, their representations still lodge and insidiously perpetuate misconceptions of sharks. James Bond and his leading lady are suspended over an indoor shark pool in Roald Dahl and Lewis Gilbert's *You Only Live Twice* (1967). In William D. Wittliff and Wolfgang Peterson's *The Perfect Storm* (2000), the crew suddenly find that the fish they've just landed is not a swordfish but a very large, angry mako, half awash on the deck and certain to bite someone: a necessary cameo part for a shark in any potential shipwreck scenario. The events of this well-made disaster movie have nothing to do with sharks, only with hurricane winds and hundred-foot waves, but the mako contributes the essential attribute of sea-life that comes with giant waves. Crabs, jellyfish and seaweed, even plastic bags and bottles – the human contribution to the seascape – cannot compete with the dramatic impact of an angry shark on deck to denote the danger of being at sea. In John Hodge and Danny Boyle's *The Beach* (2000), the character Richard suddenly assumes the position of 'alpha male' in the island community when he manages to spear an inquisitive juvenile mako shark. More interesting than the shots of the shark is the story that he tells, to a fascinated and silent audience, of how he overcame the mighty adversary. The scene encapsulates the way in which monster shark stories are created, elaborated and accepted by society. Richard's story then takes a sudden turn when he announces that just before it died the shark told him to have a nice dinner. The absurd representation of the shark is pushed to the next level and the suspense collapses into humour.

Andrew Stanton and Lee Unkrich's *Finding Nemo* (2003) was a highly successful animation about an ocellaris clownfish, the only hatchling of the batch following a devastating attack on their nest by a barracuda. When little Nemo is taken by a diver, his father Marlin seeks help from other sea life to find him, including three sharks: a mako, a hammerhead and a great white. These might be dangerous friends to fish, except that they have sworn an oath to improve the image of sharks by becoming vegetarians. When Marlin and his friend Dory try to escape, Dory finds the diver's mask with his address etched in it and refuses to leave without it. Dory gets a nosebleed, filling Bruce's nostrils with fish blood, propelling the brutish Bruce's instincts to overwhelm his best intentions, and they narrowly escape being eaten. Although the scene is amusing, its representation of sharks is as crazy, toothsome creatures that are ultimately hostage to their own hunting instincts. In spite of Bruce's stated intentions to reform the image of sharks, the movie merely contributes to their negative representation.

While *Open Water* (2003), written and directed by Chris Kentis, is a serious movie and unusually accurate in its depiction of sharks, the film is essentially about a distressed marriage and the danger of existential abandonment. However, it makes excellent use of shark fins surfacing, circling and vanishing, creating a looming sense of threat more convincingly than any of the major shark-focused movies. Establishing threat from external forces is crucial in the creation of audience fear (Reyes, 2016). The premise of the picture is an actual event: two young married divers were left behind by their dive boat at the Great Barrier Reef. Some of their equipment was later found, but their bodies were never recovered. *Open Water*'s director and producer are also a married couple, as well as divers themselves, and therefore aware of the patterns of domestic strife along with the risks of being abandoned at sea, risks that are presented without exaggeration or inaccuracies. The result is a persuasive natural tension that mounts without contrivances. The young couple continue to argue and blame one another as sharks gather, mainly below the divers' legs. The sharks are real and really swimming around the actors for hours at a time. However, these sharks are barely six feet in length and don't charge, they merely brush against the discernibly weakening divers with increasing frequency and interest. Eventually, in a fictional moment, one shark ventures a trial bite. That seems like the beginning of the expected gory end, but the movie opts instead for realism. Caribbean reef sharks are generally non-aggressive, and we can testify to their being surprisingly accommodating to divers in their environment. The mortal dangers for anyone lost at sea are hypothermia and growing fear, effects that in this movie become intolerable. The sharks are used dramatically to increase the tension that eventually reunites the bickering couple – they begin to support each other emotionally as well as physically. In an intimate gesture that serves as a counterpoint to their separateness established in Act I, Daniel holds Susan's hair as she vomits from seasickness. When he dies in the night, either from hypothermia or blood loss or a combination of the two, Susan lets his body drift away, then turns from the sight of the sharks closing in on his carcass. Unable to withstand her ordeal alone, she drowns herself. The decision not to sensationalise the sharks in the end worked well for what was actually a suspense thriller, rather than a shark movie.

In Petter Skavian, Joachim Ronning and Espen Sandberg's *Kon-Tiki* (2012), based on the book of the same name, a shark's fin is spied following the raft. Such a sighting is accurate enough. During months at sea, sailors enjoy sightings of whales, albatrosses, flying fish, phosphorescence, and curious sharks. In Act II, a curious great white shark is spotted circling the makeshift basket shark cage suspended from the raft. One sighting of the cruising shark causes panic.

Vessels trailing odours from garbage, cargo or excrement are followed by the sharks – human discharge is also 'chum'. When the raft's resident macaw, Lorita, tries to alert them to a shark, they appear not to hear it. Lorita goes for a swim and sacrifices itself to demonstrate a point to the sailors – who now give her their full attention. The shark is hauled out of the ocean, its jaws snapping, while close-up shots reveal lower legs of sailors standing dangerously close to the teeth, providing an easy thrill for the audience. The shark is stabbed repeatedly in the head. Its underside is then sliced open in a desperate attempt to retrieve the parrot, with the consequence that more sharks are now attracted to the raft by the smell of fish blood in the water. There is no indication that the shark would have attacked the sailors. It is made explicit that the big attraction for the other sharks that now surround the disintegrating wooden raft is the sushi the sailors have prepared for them. Herman, devastated by the loss of his parrot, the accumulating sharks and the fact that his suggestion of using steel cord to secure the raft has been abandoned, along with the means to do so, then slips into the ocean. We see him from the perspectives of the sharks, which swim away. More fish guts are thrown to distract the sharks, but these great whites show no apparent interest in Herman or a second sailor who dives in to help him back to the raft. However, as Herman's feet make it onto the raft, a great white makes a close investigatory pass. No one is eaten by sharks, either in the movie or during the Kon-Tiki journey on which the book and movie were based. Most of us would be surprised to learn that interest without predation is more usual shark behaviour. Recent research in the field of ocean geography has revealed that many shark encounters go unreported officially, precisely because they are positive experiences and not newsworthy human drama (Gibbs and Warren, 2014).

Shark documentaries

While the movie business has a clear agenda of thrilling and entertaining people with visual spectacles, other moviemakers, often with conservation interests, are more interested in providing information about sharks to correct gross misunderstandings and change attitudes towards sharks.

However, even filmmakers with the best intentions can sometimes undermine their efforts by failing to understand the importance of the musical score that accompanies their verbal and visual narratives in establishing mood and attitude. While no serious shark documentary filmmaker would be likely to choose John Williams' *Jaws* score while attempting a truthful depiction of the shark, there are more subtle ways in which the score can work against an objective agenda. A recent study asks even documentary filmmakers to think more carefully about the music they use to accompany their shark footage. In the first study of its kind, Andrew Nosel argues: 'given that nature documentaries are often regarded as objective and authoritative sources of information, it is critical that documentary filmmakers and viewers are aware of how the soundtrack can affect the interpretation of the educational content' (Nosel, Keenan et al., 2016).

Peter Gimbel and James Lipscomb's *Blue Water White Death* (1971) suffers from no manipulative soundtrack. The film remains one of the best and most interesting shark documentaries, and it holds a special place in film history. The filmmakers assembled a crew that included the renowned photographers Ron and Valerie Taylor, the amazing underwater cinematographer Stan Waterman, the author Peter Matthiessen, and (serendipitously) the folk singer Tom Chapin on a voyage from South Africa to Sri Lanka to Australia in search of what proves to be the elusive great white shark. The result is an odyssey fully documented on film. Why were white sharks difficult to find in South Africa and in the Indian Ocean? Probably because the filmmakers didn't yet know where to look. The film was shot almost at the dawn of time in terms of our knowledge of sharks. 'A feeding frenzy' was believed to be just that, a chaos of sharks biting and tearing at everything and anything. And yet this crew jumps into the water, at night, to film the action around a dead whale beset by sharks. The footage is stunning. The divers push away sharks or fend them off with their cameras. Their courage seems unimaginable. They dare to correct the universally held belief in the 'feeding frenzy', discovering that feeding sharks are no more frenzied than diners at a busy buffet. Eventually, after more than six months of searching, Gimbel recruits Rodney Fox, the poster boy for shark attack, to guide the filming in Spencer Gulf, and the team is rewarded with iconic sequences of huge white sharks swimming with their jaws open. But what remain most memorable are the film's unscripted moments, both on board and in the water, as the players interact authentically.

The more recent *Sharkwater* (2006), written and directed by Rob Stewart, approaches sharks from an environmentalist and political perspective, documenting some of the slaughter that amounts to an estimated 100 million sharks per year, primarily for shark fin soup. Canadian journalist and biologist Rob Stewart travels to Costa Rica and the Galapagos to feed sharks by hand and thus counter popular misapprehensions about sharks and demonstrate how little danger they actually present. The places that Stewart chooses coincide with the favourite haunts of shark poachers, and he teams up with Paul Watson of Sea Shepherd fame to film the illegal shark-fin trade. *Sharkwater* is visually stunning, filmed in high-definition video. Whether it's a more effective defence of sharks because of the beauty of the photography or a more substantive, beautiful film for its strong environmentalist principles is a question we do not need to answer. The film makes its case with a powerful combination of visual pleasures and conservationist principles. Sadly, Rob Stewart was working on a sequel, *Sharkwater: Extinction*, when he died in a diving accident while learning to use a re-breather at a depth of 200 feet. He was only 37 years old.

This is Your Ocean: Sharks, written and directed by George Schellenger, is a documentary film about Jim Abernethy and his attempt to reach a wider audience to share his appreciation of tiger sharks through exposure to them in the Bahamas, and through shark art in a wide range of contexts. His approach is worthy and it appears to work well, especially when his friendly artists take their work into classrooms. However, the documentary is somewhat marred by Abernethy's inspirational soundtrack, and especially by his over-reliance on the submissive charm of Emma the tiger shark. However close divers may feel to sharks, it is not a good idea to assume reciprocity and take risks, in front of the camera or otherwise, with wild creatures. Abernethy's footage, which features him lying on the bottom of the ocean waiting for Emma to approach his head, is particularly worrying, however well meant as a reassurance to viewers that these are not mindless man-eaters. It only takes one mistake to undo all of the great work people like Abernethy are doing in popularising sharks. Sharks should be free to be admired from a distance, without petting, massaging them into stupors, or worse – making them underwater vehicles, or window-dressing them as furless dogs. As every diver knows, sharks have their own personalities and they all need to be treated with respect, which includes being mindful of their need for a little distance, even when the camera shot seems just too good to miss.

In 2015 the BBC launched a documentary series called *Shark*. This series of four programmes directed by Steve Greenwood was useful, well balanced and objective. It took cameras into the oceans to look at sharks, from a distance, in order to learn more about how they coped with the challenges that modern human existence is presenting them with, whether it be boat propellers, pollution or shark-fin hunting. The visual representations of the sharks were mediated by factual narration that was suitably inquisitive and informational without overloading the general viewer with scientific language. Even in this otherwise superlative production, the soundtrack often exaggerates the noise of sharks moving, opening their mouths and biting, against a rapid guitar and drum score – at times to the point of sensationalism. The final programme in the series took the viewer behind the scenes in order to demonstrate the extreme patience and understanding required to be able to deliver such excellent footage of these often shy creatures. The scenes that no doubt gripped most viewers were those that featured the divers in the ocean with the great white shark. The shark's behaviour was being monitored carefully throughout the filming and explained clearly to the viewer. As soon as the pectoral fins were lowered in a threat display, showing the black spots on each fin, the diver was shown moving away carefully, without distressing the shark. The documentary filmmakers announced at the end of the series that in all of the extensive filming of sharks during the course of making the programme there was not one negative encounter, and it was apparent that one reason was the lack of provocation on the part of the crew. This documentary series was a highly positive representation that should have instilled a little respect for sharks and their daily challenges in the open oceans.

Among the reasons for film's fixation on sharks is their visual power. When a shark appears, either in film or in life, we don't look away. The cleverest comic directors understand that sharks are natural scene-stealers, and these filmmakers play up sharks' size and toothiness to the point of parody; they mock the silly, exploitative movies. Other filmmakers, particularly documentary filmmakers, also seem to be evolving toward less exploitative and more accurate presentations of sharks. Unfortunately, media such as TV news coverage, online social media and online news never tire of showing footage of shark attacks posted to social media sites. This violent footage draws enthusiastic popular responses, suggesting a dangerous level of negative emotion towards sharks. The result is that many more people are likely to access news reports about attacks than about conservation, and even fewer seek out research articles, which are often written with other experts in the disciplinary field in mind (Nghiem, Le T.P., Papworth, S.K. et al., 2016). Video plays to the short attention span. Even when research is cited accurately in news media, there often isn't the space to delve into the important subtleties of research findings. It is unfortunate that the media controls to such a large extent how we understand our relationship to sharks. Only when sharks have attacked someone do they get our attention, at which point Perth suddenly seems close to Europe or New York in the media imagination and people can finally locate Reunion Island on the map. In fact, shark attacks are extremely rare, although reported far more frequently than, for instance, much more frequent accidental deaths by drowning across the world. At the very least, we need to be more aware of rhetorical filmic, artistic strategies, tropes and schemes used for creating emotion so that we can understand that the sharks we think we know are cultural constructs created for our terror and our pleasure (Figure 7.2).

References

Audissino, E. (2014) *John Williams's Film Music: Jaws, Star Wars, Raiders of the Lost Ark, and the Return of the Classical Hollywood Music Style.* Wisconsin: Wisconsin University Press.

Brophy, P. (1986) Horrality – The Textuality of Contemporary Horror Films. *Screen.* 27.1:2–13.

Catlin, J., Hughes, M., Jones, T. and Jones, R. (2014) White Sharks in Western Australia: Threat or Opportunity? *Journal of Ecotourism.* 13.2–3, 159–169.

Connor, J.D. (2015) *The Studios After the Studios.* Stanford California: Stanford University Press.

Francis, B. (2012) Before and After 'Jaws': Changing Representations of Shark Attacks. *The Great Circle.* 34.2, 44–64.

Gibbs, L. and Warren, A. (2014) *Cultures and Politics of Encounter and the Sea. Australian Geographer.* 45.2, 101–107.

Grierson, T. (2016) *The Shallows*: Review. *Screen International.* June 23.

Healy, S. (2011) Film Reopens Painful Memories for Shark Attack Victim. *The Sunday Mail* (Qld).

Koldau, L.M. (2010) Of Submarines and Sharks: Musical Settings of a Silent Menace. *Horror Studies.* 1.1, 89–110.

Mulvey, L. (1989; 2009) *Visual and Other Pleasures.* Basingstoke: Palgrave Macmillan.

Muter, B.A., Gore, M.L., Gledhill, K.S., Lamont, C. and Huveneers, C. (2013) Australian and U.S. News Media Portrayal of Sharks and their Conservation. *Conservation Biology.* 27.1, 187–196.

Myrick, J.G. and Evans, S. D (2014) Do PSA's Take a Bite Out of *Shark Week*? The Effects of Juxtaposing Environmental Messages with Violent Images of Shark Attacks. *Science Communication*, 36.5, 544–569.

Neff, C. (2015) The Jaws Effect: How Movie Narratives are Used to Influence Policy Responses to Shark Bites in Western Australia. *Australian Journal of Political Science.* 50.1, 114–127.

Nghiem, Le T.P., Papworth, S.K., Lim, F.K.S. and Carrasco, L.R. (2016) Analysis of the Capacity of Google Trends to Measure Interest in Conservation Topics and the Role of Online News. *PLoS ONE.*11.3, 1–12.

Nosel, A.P., Keena, E.A., Hastings, P.A. and Gneezy, A. (2016) The Effect of Background Music in Shark Documentaries on Viewers' Perceptions of Sharks. *PLoS ONE*, Aug 3.

Reyes, X.A. (2016) *Horror Film and Affect: Towards a Corporeal Model of Viewership.* Oxford and New York: Taylor and Francis.

Wasser, F. (2013) *Steven Spielberg's America.* Cambridge: Polity.

Index

For Product Safety Concerns and Information please contact our EU representative GPSR@taylorandfrancis.com
Taylor & Francis Verlag GmbH, Kaufingerstraße 24, 80331 München, Germany

www.ingramcontent.com/pod-product-compliance
Ingram Content Group UK Ltd.
Pitfield, Milton Keynes, MK11 3LW, UK
UKHW020003200926
475702UK00015B/2458
9780367855833